Pride of the Valley

Pride of the Valley

Sifting through the History of the Mount Healthy Mill

by

Tracy Lawson

with the assistance of

Steve Hagaman

The McDonald & Woodward Publishing Company
Newark, Ohio

The McDonald & Woodward Publishing Company
Newark, Ohio *www.mwpubco.com*

Pride of the Valley: Sifting through the History of the Mount Healthy Mill

Printed in the United States of America by McNaughton & Gunn, Inc., Saline, Michigan, on paper that meets the minimum requirements of permanence for printed library materials.

First printing June 2017
10 9 8 7 6 5 4 3 2 1
25 24 23 22 21 20 19 18 17

Library of Congress Cataloging-in-Publication Data

Names: Lawson, Tracy, 1966- author. | Hagaman, Steve, 1964- author.
Title: Pride of the valley : sifting through the history of the Mount Healthy Mill / by Tracy Lawson and Steve Hagaman.
Description: Newark, Ohio : The McDonald & Woodward Publishing Company, [2017] | Includes bibliographical references and index.
Identifiers: LCCN 2017019460 | ISBN 9781935778387 (pbk. : alk. paper)
Subjects: LCSH: Mount Healthy Mill (Mount Healthy, Ohio)—History. | Flour mills—Ohio—Mount Healthy—History. | Mills and mill-work—Ohio—Mount Healthy—History. | Mount Healthy (Ohio)—History—19th century. | Mount Healthy (Ohio)—History—20th century.
Classification: LCC TS2135.U62 O36 2017 | DDC 664/.720977177—dc23
LC record available at https://lccn.loc.gov/2017019460

Contents

Dedication

Steve:
To Jordan, Sidney, Cayley, and Jakob

and

Tracy:
To my dad and uncle, Todd and Gary Stone

Acknowledgments

Dave Huser of the Mount Healthy Historical Society provided indispensible help during our research. We appreciate his willingness to comb the Historical Society's archives and files for us many times over.

Patrick Brown of Brown Studios Photography in Hamilton, Ohio, graciously provided scans of the stereoscope, which is believed to be the oldest photograph of the mill.

Dan Shaw, Operations North District Superintendent at Great Parks of Hamilton County, answered many questions and gave us access to the Park District's files relating to the mill, the fire, and its aftermath.

Walter Rogers and Craig Rogers, who shared photos and family documents, and especially Craig, who served as a sounding board and cheerleader throughout the writing of this book.

Winona Rogers Brigode and the late Florence Rogers Rieck, Tracy's great-aunts, answered her many questions and shared photos and other family documents.

Sandra Chadwick Mussey, descendant of Wallace Chadwick, who helped fill in the gaps in Mary Jane Chadwick Rogers' story.

Jerry McDonald of McDonald & Woodward Publishing, whose vision for *Fips, Bots, Doggeries, and More* has continued with this companion book, *Pride of the Valley*.

Additional help was provided by Karen Arnett, Mt. Healthy Renaissance, Mount Healthy, OH; Lesley Bell, personal trainer, Fit 180, Dallas, TX; Jim Blount, historian, Butler County, OH; Jack Bredenfoerder, First Vice President, Cincinnati Chapter, Sons of the American Revolution, Cincinnati, OH; Scott Byerly of First Edition Books, Delaware, OH; Marlene Carmack and Kathy Creighton of the Butler County Historical Society, Hamilton, OH; Joanne Good of Glendale Heritage Preservation, Glendale, OH; Brent Gregory, who shared his photos of the covered bridge and mill, Mount Healthy, OH; Elaine Temming, the late Miriam Borchelt, Joan Groff, and Patricia Buck, daughters of Ralph Groff, Cincinnati, OH; Kevin and Cindy Hardwick, current owners of the Jediah Hill homestead, Mount Healthy, OH; Millie Hartman, granddaughter of Charles Hartmann, Mount Heathy, OH; Leland Hite, steam historian and author, Loveland, OH; Thomas Jorstad of the Smithsonian Institution, Washington, DC; Kristin Kitchen, owner of Six Acres Bed and Breakfast, College Hill, OH; Sienna Logan of the Hartford Historical Society, Hartford, CT; Bob Mason of Great Parks of Hamilton County, Cincinnati, OH; Robert McMaken, genealogist, Butler County Courthouse, Hamilton, OH; Andrea Neço, civil structural engineer, Milan, Italy; Carrie Kettell Parrett, daughter of late historian Carolyn Kettell, Cincinnati, OH; Don Prout of CincinnatiImages.com, Cincinnati, OH; Larry Pyle, architect and local historian, Cincinnati, OH; Robert T. Rhode, steam historian and author, Springboro, OH; Sharonville United Methodist Church, Sharonville, OH; Timeless Photo Restoration, Downers Grove, IL; James Wolf, mayor of Mount Healthy and great-great-grandson of Charles Hartmann, Mount Healthy, OH: Vermont Timber Works, North Springfield, VT; all of whom we thank.

Pride of the Valley

Sifting through the History of the Mount Healthy Mill

Introduction

This little piece of land in Ohio says so much about the history of our country. First it was woodlands inhabited by the Native Americans. The early settlers processed the lumber as they cleared the land, and then later needed a flour mill as Ohio shifted to an agrarian society. [The mill's] demise happened over time, until it became a "parking" spot and an outmoded symbol of Americana. [It's] like Roots from a geographical perspective, with the drama that comes from changing times.

— Tom Jesionowski, great-grandson of Mount Healthy Mill owner Charles Hartmann, from email correspondence with the author

I am an amateur historian and genealogist, and *Pride of the Valley* is a book about my search for the history of the Mount Healthy Mill (Figure I.1). The mill, which was built by my great-great-great-great grandfather, Jediah Hill, stood on the banks of the West Fork of Mill Creek in Springfield Township, Hamilton County, Ohio, from the 1820s until 1981. The mill was a well-known landmark in the area, but much of the history of the structure and the people who had created, operated, maintained, and nurtured the business had, over time, slipped into obscurity.

Figure I.1. This 1907 photograph likely depicts a Sunday afternoon at the Charles Hartmann home. The Mount Healthy Mill is the large white structure with the red roof.

Jediah Hill opened his sawmill in the 1820s, and brought Henry Rogers, Jr., into the business as a hired hand and, soon thereafter, into the family as his son-in-law. The mill continued to be owned and operated by my ancestors until 1883, at which time it was sold out of the family. Successive owners continued to run the mill until 1952, when the land on which it stood was procured by the Army Corps of Engineers. The mill, once vacant, fell victim to vandals and the ravages of nature, yet it remained standing until October, 1981, at which time it was destroyed by a fire of suspicious origin.

If I had developed my current interest in the mill as a teenager, while living a mere fifteen miles away in Sharonville, Ohio, and had visited the site before the fire, all I would have seen was the hulking, abandoned shell of the building, without being able to recognize it as the remains of a once-thriving business. I might have hesitated to approach the structure because of the foreboding look of the broken and boarded-up windows. I would have neglected to notice and admire my ancestors' carpentry skills. I wouldn't have been able to appreciate their ingenuity and understand that they had built a water-powered factory — and did so without using power tools.

With time, however, my interest in my family's history grew, inspired and facilitated in no small part by a journal kept in 1838 by my great-great-great grandfather, Henry Rogers. That journal was the record of a trip taken by Henry, his wife Rachel, and her parents, Jediah and Eliza Hill, and when I received a typewritten copy of it, it drew me into their time and their lives.

Years of studying that document led me to write my first book, *Fips, Bots, Doggeries, and More*, which was published in 2012. In *Fips, Bots, Doggeries, and More,* I shared an annotated version of the journal that chronicles the family's 1838 horse-and-wagon trip from the village of Mount Pleasant (now Mount Healthy), located twelve miles north of Cincinnati, to New York City. They traveled at a leisurely pace, touring places of interest and allowing time for visits to family and friends in Ohio, Pennsylvania, and New Jersey. One important purpose of that trip, however, was to allow Jediah and Henry to observe working mills and get ideas for expanding their own sawmill so that it could grind grain too.

The Mount Healthy Mill received no more than passing attention in my first book. I had focused my research on the journey, and beyond grasping the mechanics of a typical nineteenth-century water-powered mill, I put limited effort into learning more about the structure, its mechanics, its business, and the greater family that created and maintained it.

Now, however, my focus has shifted. My coauthor, Steve Hagaman, and I have taken on the task of working backward through the nearly two centuries since the mill's wheel began to turn, and re-building the mill through all its phases, from birth to death. Here we explain the history of the mill and, in the process, we'll share our insights as to why the mill was important enough to get excited about in the first place.

To understand the greater history of the mill, we must also know the stories of generations of pioneers, farmers, businessmen, inventors, immigrants, community leaders, and families whose individual contributions to the building and the evolution of the original, primitive sawmill on the Ohio frontier enabled it, for nearly one hundred and thirty years, to meet the changing needs of a growing community.

The facts presented in *Pride of the Valley* are infused with my notes and observations as they related to the research process, all of which I hope will provide a humanizing touch to my work and encourage other individuals to take on similar research projects that relate to their own families.

What Happened Next?

I've studied and written about Henry Rogers' journal for over two decades, and I know my great-great-great grandfather as more than just a headstone in a cemetery plot or a name on a family tree. He was a skilled carpenter, a businessman, and an inventor. He was smart and funny and interested in just about everything. I wish he had left behind more than just one journal because, after *Fips, Bots, Doggeries, and More* was published, I was left with an appetite for additional parts of the story. There was still so much I didn't know.

Henry's journal, however, ends abruptly upon the family's arrival at Mr. James Townsend's residence at 707 Greenwich Street in Manhattan,[1] with the notation "End of Volume One." There is no evidence that Volume Two ever existed.

I couldn't blame Henry if he'd failed to take an interest in penning daily entries on the return trip, but the lack of a defined ending to the story left me with many unanswered questions. Chief among them was this: did Jediah and Henry arrive home with a plan for expanding their business? If so, when did they begin the enormous structural changes that would have been necessary to add flour milling to the enterprise?

My first attempts to answer that question were disappointing. The National Register of Historic Places Nomination Form for the Mount Healthy Mill, prepared in July, 1980, reads in the Statement of Significance:

> "Following [Jediah] Hill's ownership and operation, the mill was owned by Henry Rogers and then by his son, Wilson Rogers, who apparently continued operation as a saw mill. Around 1887, the mill became the property of Charles Hartman[n]. As timber in the area became scarce, Mr. Hartman[n] also began to mill coarse flour for nearby farmers for home use." [2]

The land on which the mill once stood is now part of Winton Woods, one of the Great Parks of Hamilton County. Park officials shared their files, which recount the park's efforts to preserve the vacant, run-down mill in the late 1970s. Interviews with Ralph Groff, the mill's final owner, and with former Mount Healthy Mayor Albert Wolf, a grandson of Charles Hartmann, seemed to confirm what I'd read in the available local histories.

Apparently, three generations of my family had operated a sawmill.

But that couldn't be right. I didn't want to believe that Jediah and Henry went to the trouble of taking that trip back east, only to give up on their plan to expand the mill.

I found a glimmer of hope for getting more information in the Great Parks of Hamilton County files. A photocopy of an 1869 Hamilton County map showed a "G & S Mill," which I took to mean "Grain and Saw Mill," on Henry Rogers' land. But, when I interviewed Carolyn Kettell, a local historian, longtime resident of the area, and one-time owner of the Jediah Hill homestead, I asked her about the map, and she told me the notation had to be a mistake.

Shouldn't I be able to trust a map drawn during the time period in question? It would be a long time before I had the answer.

According to the available records, German immigrant Charles Hartmann, who purchased the mill from Wilson Rogers in 1883, was credited with converting it from a sawmill to a flourmill, and did so only because timber in the area had become scarce. He was also responsible for removing the original water wheel and upgrading to steam power in 1898.

Hartmann sold the mill to C. C. Groff in 1911, who later converted it from steam to run on a diesel engine. Members of the Groff family owned and operated the mill until it closed in 1952. These last two upgrades ushered the business into the twentieth century, and the mill's capacity grew within its expanded walls until, when called upon, it could produce fifteen thousand pounds of flour a day to supply restaurants, hotels, bakeries, and homes throughout the Cincinnati area.[3]

What do you imagine Jediah and Henry would've thought of that? I'm sure they would have been amazed. All the same, I couldn't bear to think of the possibility that their achievements were being misreported, ignored, or deleted from the record of the mill's history.

So, I set out to explore the story in greater depth. I had honed my research skills while writing *Fips, Bots, Doggeries, and More*. I was ready to delve into the lives of my ancestors and the subsequent owners of the mill, and excited to attempt to unravel the details of the mill's history, but I would have had no idea how to explain, let alone understand, the mill's structural evolution without the expertise of my coauthor, Steve Hagaman.

Steve had developed his own interest in the Mount Healthy Mill and had enjoyed exploring the ruins that lie on the banks of Mill Creek. As a result, he got involved with the Mount Healthy Historical Society, read *Fips, Bots, Doggeries, and More*, and contacted me on Facebook to see if I had any information that would help him with his own research.

We were both thrilled to communicate with someone who shared our interest (though Steve uses the word obsession) in the Jediah Hill mill. Steve's training as a carpenter and historic millwright gave him insight and knowledge that I lacked. We passed information to one another, but carried on our pursuits separately. My book outline centered around the mill's influence on the community and profiles of the families who owned it. Steve, meanwhile, was busy reconstructing the building in his imagination.

In the spring of 2014, Steve had a breakthrough. He was sure he'd determined the location and path of the original mill race, and offered to show it to me next time I was in town. In August of that year, my dad, my brother Greg, my niece Reagan, and I ended up traipsing around in the woods and Mill Creek with Steve as he explained just how much work and planning had gone into building the original mill (Figure I.2).

About halfway through the afternoon, we were jumping from rock to rock in the creek and it occurred to me to ask Steve if he'd like to contribute his knowledge to my book. Just like that, we became coauthors.

Even though our efforts to locate business ledgers or millers' journals came to naught, we did find clues in land and census records, a poem, and a stereoscope image from the 1860s. That might not sound like much, but it was enough to go on, and those details directed us to other long-forgotten information that collectively supports a logical progression of events. While we both spent time dredging up information in courthouses and libraries and poring over digitized books and records on the internet, Steve also continued to play in the creek, searching for evidence of the early dams and checking out the intake for the millrace. He was a lot braver than I about digging around in the ruins of the mill site!

Figure I.2. Todd Stone, a great-great-great grandson of Jediah Hill, stands in the ruins of the doorway to the original sawmill.

Our combined findings yielded complementary information and also challenged several details accepted as facts in the then-current account of the mill's history. Now, with *Pride of the Valley*, we're excited to share our findings and, in the process, update the history of the Mount Healthy Mill.

Meet the Families

Members of five generations of my ancestors and those of the Hartmann and Groff families, all owners of the mill during part of its history (Table I.1), are mentioned in this book. To orient you, the reader, and to help you keep all those names and relationships straight, we have included diagrams showing the genealogical relationships of members of each family discussed in the book.

The first of these genealogical diagrams and a corresponding list of names of the family members included in that diagram appear here (Figure I.3). It was difficult to get everyone in the Hill and Rogers family tree on one page, but the master

Table I.1. Owners of the Mount Healthy Mill

Owner	Years the Mill Was Owned	Number of Years Owned
Jediah Hill (1793–1859)	c. 1820–1859	c. 39
Henry Rogers, Jr. (1806–1896)	1859–1875	16
Wilson Rogers (1843–1927)	1875–1883	8
Charles Hartmann (1844–1918)	1883–1911	28
Claude C. Groff (1867–1936)	1911–1936	26
Ralph Groff (1893–1984)	1936–1952	16

chart on page 9 shows, at a glance, the number of children in each successive generation and how the two lines connect with the union of Henry Rogers, Jr., and Rachel Maria Hill. Smaller, more detailed descendant charts preface eight chapters in Section I of the book that pertain to each successive generation of the families that owned the mill.

Pride of the Valley tells the story of the beginning, life, and eventual demise of the Mount Healthy Mill, which operated on the banks of the West Fork of Mill Creek for one hundred and thirty years. Let's begin, in Section I, by getting to know the families who owned the mill and examining how their lives were interwoven with each other as well as with pivotal events in our country's history. We will follow that exploration, in Section II, by looking more closely at the architectural and mechanical evolution of the building itself and the legacy of the mill once it was retired.

Finally, the Appendixes provide (I) a detailed timeline of the mill and the people who built, owned, and operated it, (II) a detailed site plan of the mill, and (III) a photographic record of the mill, and an extensive Bibliography identifies published reports and other sources from which so much of the information in the book was derived.

Figure I.3. Five generations of the Hill and Rogers families that were involved in the Mount Healthy Mill are represented in the diagram on the facing page. The alphanumeric entries in the diagram represent the following members of the two families: **Row 1:** 1AA - Paul Hill 1751–1824, 1AA+ - Rachel Stout 1760–1820; **Row 2:** 2A - Henry Rogers, Sr. 1752–1840, 2A+ - Phoebe Burnet 1758–1812, 2AA - Samuel Hill 1785–1827, 2AA+ - Mary Woolverton 1781–1838, 2BB - Benjamin Hill 1787–1844, 2BB+ - Margaret Vandike 1790–1850, 2CC - Charles Hill 1790–1855, 2CC+ - Mercy Hendrickson 1795–1869, 2DD - Jediah Hill 1793–1859, 2DD+ - Eliza Hendrickson 1797–1854, 2EE - David Hill 1795–

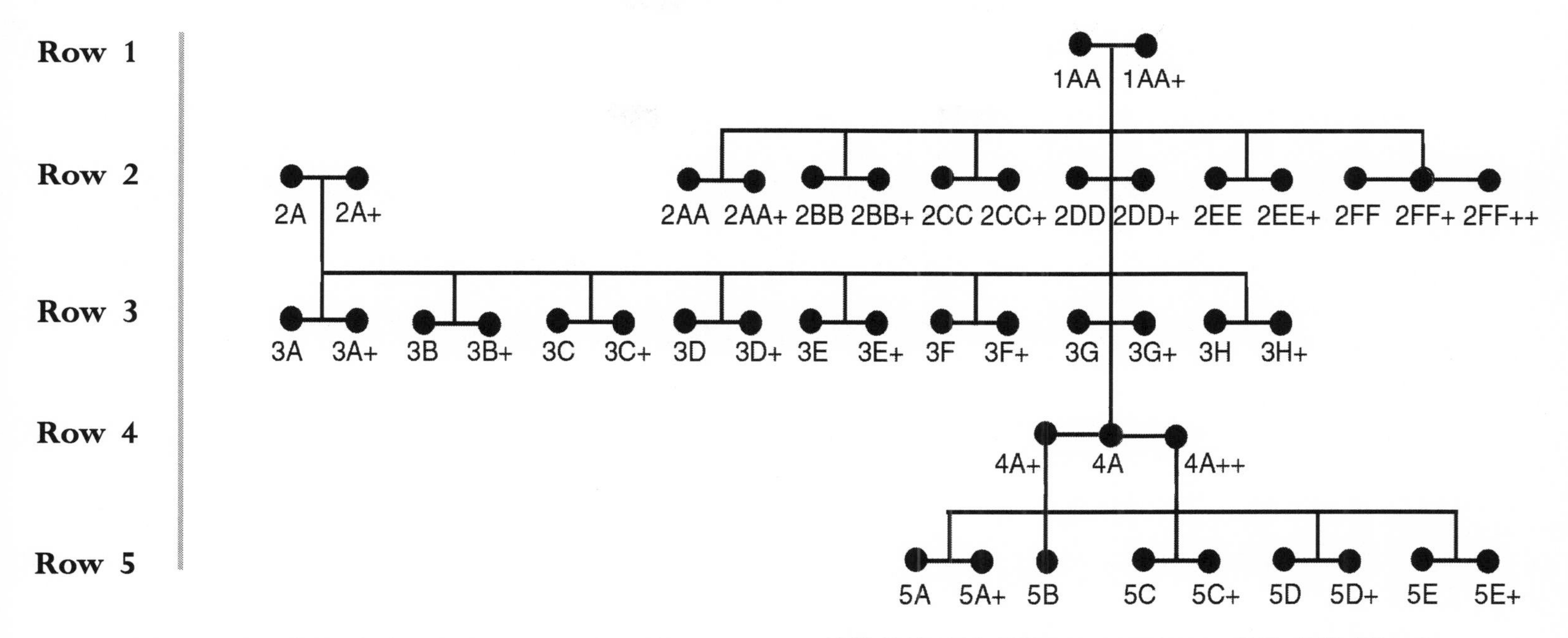

1865, 2EE+ - Mariah Hendrickson 1789–?, 2FF - Asher Hill 1798–1879, 2FF+ - Rachael Green 1789–?, 2FF++ - Margaret Green 1801–?; **Row 3:** 3A - Elizabeth Rogers 1786–1875, 3A+ - Thomas McFeely 1780–1859, 3B - Phoebe Rogers 1787–1833; 3B+ - Jonathan Holden 1796–1845, 3C - Sarah Rogers 1789–1863, 3C+ - Michael Burdge 1787–1861, 3D - Hannah Rogers 1791–1875, 3D+ - Zebulon Strong 1788–1875; 3E - Jemima Rogers 1795–?; 3E+ - Richard McFeely 1787–1837, 3F - Nancy Rogers 1802–?, 3F+ - Cyrus Brown 1799–1877, **3G - Henry Rogers, Jr. 1806–1896, 3G+ - Rachel Maria Hill (3A) 1816–1888,** 3H - Maria Rogers 1810–1869, 3H+ - Levi Pinney 1804–1839; **Row 4:** 4A - Wilson Thompson Rogers 1843–1927, 4A+ Mary Jane Chadwick 1843–1881, 4A++ - Nancy Gwaltney Rogers 1846–1917; **Row 5:** 5A - Harry Chadwick Rogers 1867–1951; 5A+ - Margaret Helms Case 1871–1947, 5B - Walter Henry Rogers 1869–1888, 5C - Pearl Blaine Rogers 1884–1971, 5C+ - Alice Rosebud Brown 1886–1955, 5D - Orpha Maria Rogers 1886–1964, 5D+ - Russell Blake 1888–1969, 5E - Jay Ferris Rogers 1889–1970, 5E+ - Laura Carroll 1888–?

Section I

The Families

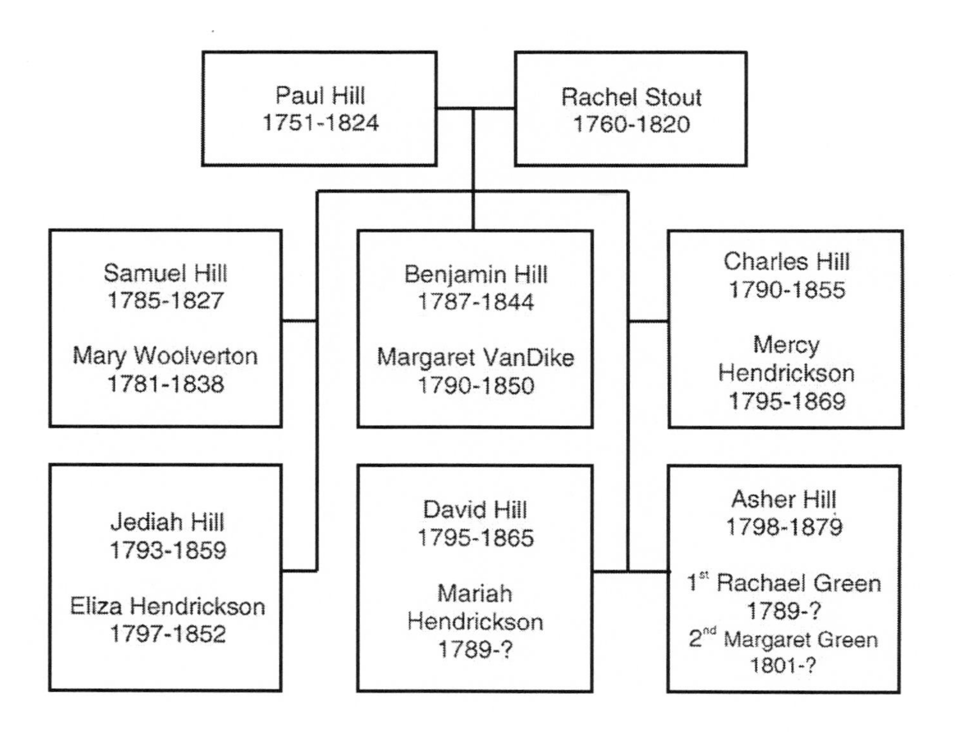
Paul Hill
1751-1824
Rachel Stout
1760-1820
Samuel Hill
1785-1827
Mary Woolverton
1781-1838
Benjamin Hill
1787-1844
Margaret VanDike
1790-1850
Charles Hill
1790-1855
Mercy
Hendrickson
1795-1869
Jediah Hill
1793-1859
Eliza Hendrickson
1797-1852
David Hill
1795-1865
Mariah
Hendrickson
1789-?
Asher Hill
1798-1879
1st Rachael Green
1789-?
2nd Margaret Green
1801-?

1
Samuel Hill and the First West

What would it take for you to leave your current job and living situation and make a long and dangerous journey to a place you'd only heard about, where opportunity might be great, but also where, at least in the short term, conditions would be more primitive and uncertain than what you'd left behind? What if there was little protection from disease and natural disasters? But what if, by going, you could test your determination, your skill, and your knowledge to mold a future for yourself and your family where, it was said, natural resources and opportunity were plentiful?

Following the Revolutionary War, the Continental Congress of the United States had no power to raise revenue by direct taxation of its citizens. With war debts to be paid, Congress adopted the Land Ordinance of 1785 on May 20 of that year. The immediate goal of the ordinance was to raise money through land sales in the new western territories.[4] This, along with land bounties given to Revolutionary War veterans, encouraged settlers to migrate to land west of the Alleghenies.

In 1793, when John Ludlow, the first recorded settler in the area that is now the village of Saint Bernard in Hamilton County, Ohio, advertised for settlers to come to the area, fear of Native American attacks understandably kept people away.[5] Most tribes did not agree with or acknowledge the United States' claim to the land, and hostilities against the white interlopers were common.[6] Two decades later, after the resolution of The War of 1812 and the Treaty of Ghent had lessened the threat to new settlers from the east,[7] interest in the Ohio lands surged, and soon people flocked to the area.

The Ordinance of 1784 called for the land west of the Appalachian Mountains, north of the Ohio River and east of the Mississippi River, to be divided into ten separate states.[8] The 1785 Land Ordinance established the Public Land Survey System, under which land was to be systematically surveyed into square townships, six miles on a side. Each of these townships was sub-divided into thirty-six sections of one square mile, or 640 acres, and could then be further subdivided for re-sale.[9]

Springfield Township in Hamilton County, where our story takes place, was part of the Symmes Purchase.[10] Many people who bought land in the Symmes Purchase hailed from New Jersey (Figure 1.1).

One such individual, named Caleb Shreve, purchased all of Section 27 in Springfield Township, but he died without ever coming to Ohio.[11]

Symmes Purchase lands had a "forfeit" provision: owners had to agree to make certain improvements within two years of making the land purchase, or forfeit one-sixth of their land to any man who had been of assistance in settling the area. This allowed individuals who lacked the finances to purchase land outright to find work as laborers, establish themselves in the community, and obtain free land through forfeit when the opportunity arose.

Caleb Shreve did not fulfill the requirement to improve his land, so he forfeited 105 2/3 acres, which was claimed by Daniel Cameron. Shreve's heirs sold the remaining 534 1/3 acres to Samuel Stout Hill in 1814.[12]

Samuel, the eldest of the six sons of Paul Hill and Rachel Stout, was born on Christmas Day in 1785. He and his five brothers grew up on the family farm in Amwell Township, Hunterdon County, east of Trenton, New Jersey. The Hills and the Stouts had large families and were well established in the area. Samuel was around twenty-seven years old when he, his wife Mary, and their two young daughters went west in 1813.[13]

Samuel and his family settled on his land, and later purchased 40 more acres in the northwest corner of adjoining Section 26 from Benjiah Cary in 1816. In 1817,

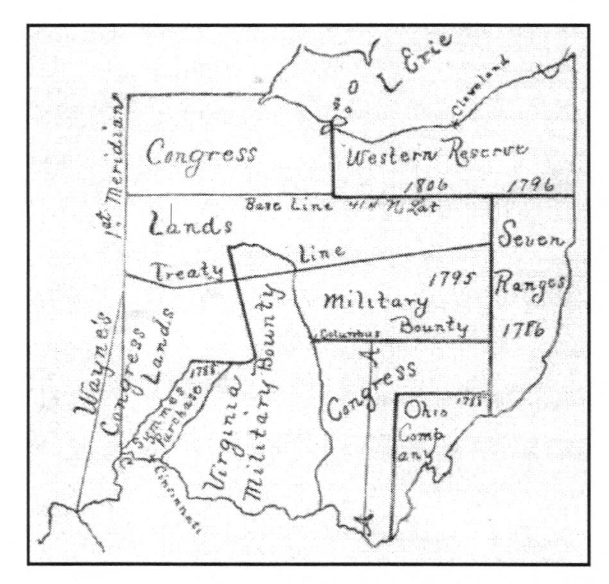

Figure 1.1. John Kilbourne (1787–1831), a writer and publisher and one of the early settlers of Worthington, Ohio, included this map, which shows the location of the Symmes Purchase as well as congressional, military, and other private company lands, in his *Ohio Gazeteer*. The *Gazeteer*, which first appeared in 1816, is recognized as Ohio's first bestseller.

Samuel Hill and John P. LaBoyteaux, his neighbor and another early settler from New Jersey,[14] co-founded the village of Mount Pleasant at the intersection of Sections 26, 27, 32, and 33.[15]

Samuel was a catalyst, without whom this story would have turned out quite differently. I imagine he envisioned a prosperous future on the new frontier. Over the next eight years, his family grew to include five daughters, Rachel, Mary, Viola, Hannah, and Margaret.[16] Samuel probably encouraged his father, Paul, to purchase land in Springfield Township, and he certainly welcomed his younger brothers, Charles and Jediah, when they brought their own families to the area.

In 1825, Samuel and his wife, Mary, sold a town lot on the east side of Perry Street, just north of Compton Road, to the five trustees of the Mount Pleasant Union Meeting House Association for a sum of twenty-five dollars. A brick structure intended for the use of all denominations to the exclusion of none was built on the lot and served the community in many capacities (Figure 1.2). Most of the Presbyterian churches in the area were formed there. Other organizations, among them the anti-slavery Liberty Party, used the meeting house for rallies and conventions. The meeting house stood on the corner of Perry Street and Compton Road for over one hundred twenty-five years, until 1966, when it was moved to the city park and restored. Currently, this structure houses the museum and archives of the Mount Healthy Historical Society.

The plans and dreams Samuel had for his and his family's future in Ohio were cut short when he died on March 11, 1827, at the age of forty-one. No will or details about the cause of death have been discovered, but my research unearthed a mystery — how did Samuel Hill purchase land in 1830, three years after his death?

Samuel Hill's Posthumous Land Purchase

The old plat of the four sections that formed the heart of Mount Pleasant indicates that Samuel Hill purchased the northeast quarter of Section 26 in 1830, but I had evidence — from more than one source — that Samuel had died in

Figure 1.2. The Free Meetinghouse, now the home of the Mount Healthy Historical Society's museum and archives, was moved to its current home in 1966.

1827. Could there be a mistake on the map? I'm inclined to believe maps, but it was hard to reconcile the date of Samuel's supposed land purchase with the death date carved on his headstone. There had to be a logical explanation (Figure 1.3).

Sections 8, 11, 26, and 29 in each township were set aside as Congressional lands, to be held and sold later, after the nearby land was settled. The government had reserved those sections in the hopes that the value would increase after other land nearby had been improved.[17] Could Samuel's posthumous purchase of the land in Section 26 somehow be connected to its designation as a Congressional Section?

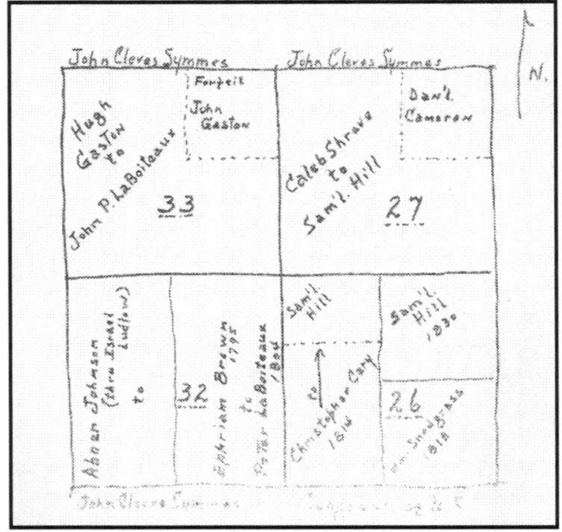

Figure 1.3. Mount Pleasant is situated in the center of Sections 26, 27, 32, and 33 of Springfield Township, and was laid out by Samuel Hill and John Laboyteaux in 1817.

The explanation lay in three deeds on file at the Hamilton County Recorder's Office. The first, found in on page 279 in volume 23 of the county deed records, was a document signed by the president of the United States, awarding a patent to Samuel Hill for the purchase of federal lands. This deed states:

> "There is granted by the United States unto the said Samuel Hill and to his heirs the quarter lot of section of land above described, to have and to hold, the said quarter lot or section of land with the appurtenances unto the said Samuel Hill and to his heirs and assigns forever. In testimony whereof I have caused these letters to be made patent and the seal of the General Land office to be hereunto affixed. Given under my hand at the City of Washington the twenty second day of October in the year of our Lord one thousand and eight hundred and twenty five, and of the Independence of the United States of America the fiftieth by the President J. Q. Adams. Geo. Graham Commissioner of the General land office".

In the margin is a hand-written notation: "Page 497 Recorded 6th April 1830."

That made sense. Samuel purchased the land in 1825, but as the deed had to pass through extra channels of government bureaucracy, it was not entered into the Hamilton County Recorder's books for four and a half years.

After Samuel died, his widow, Mary, and his brothers, Charles and Jediah, settled his estate. Mary made arrangements to sell 107 acres of Section 26 to John West, and another 55½ acres to John Larowe. Both deeds of sale were dated January 2, 1829.

The John Larowe deed was recorded February 12, 1830. The John West deed and Samuel Hill's 1825 patent for the land were both recorded on April 6, 1830. By selling the land, Mary may have prompted the original patent to finally be recorded in Hamilton County.

Figure 1.4. Drawing of John Cleves Symmes, by Henry Howe, author of *Historical Collections of Ohio.*

SIDEBAR 1.1

Jersey Boys

John Cleves Symmes (Figure 1.4) was an attorney who represented New Jersey in the Continental Congress of 1785–1786, and organized investors into the Miami Purchase Association. In 1788, he acquired more than 330,000 acres, or 520 square miles, of land in what are now Hamilton, Butler, and Warren counties in the state of Ohio, which is known as Symmes' Purchase, or the Miami Purchase. Its northern boundary runs through Butler and Warren Counties, about twenty-five miles north of the Ohio River. It is bounded on the west by the Great Miami River, by the Ohio River on the south, and by the Little Miami River on the east.

John Cleves Symmes' marketing efforts attracted an impressive number of New Jersey residents to the Miami Purchase. Many of the surnames sounded familiar to me, and I realized that they were names I'd researched while writing *Fips, Bots* . . . The Williamsons, Comptons, Hendricksons, Hunts, Tituses, Boggses, Youngs, Harts, and many more New Jersey families saw members off to Springfield Township, Hamilton County, Ohio, in the first half of the nineteenth century.

SIDEBAR 1.2

Where Did You Come From, Where Did You Go?

The 1850 Federal Census was the first to list the name of every family member, not just heads of households. It also listed where each member of the family was born, and that makes it an excellent resource for tracking migration patterns and estimating when families moved from one state to another by the birthplaces of their children. In the 1850 Federal Census of Springfield Township, eighteen percent of the heads of households hailed from New Jersey.[18]

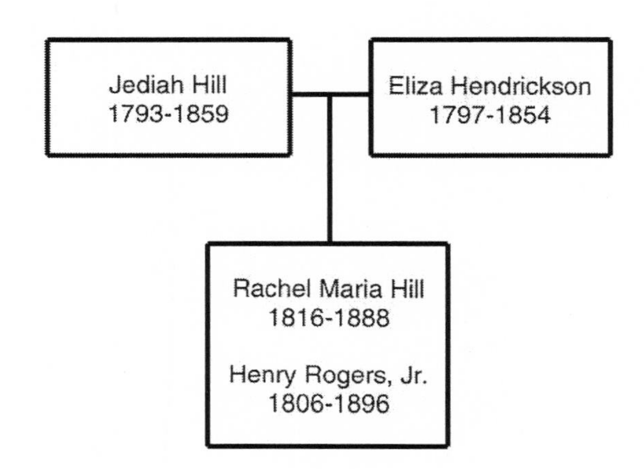
Jediah Hill
1793-1859
Eliza Hendrickson
1797-1854
Rachel Maria Hill
1816-1888
Henry Rogers, Jr.
1806-1896

2
Jediah Hill's Mill

Here's a town just beginning, like a beansprout pushing up when the ground gets warm enough for planting! A part of the country gets settled, and after that somebody starts to think about a mill. . . Then there'll be a store and a tavern and a blacksmith shop and a church and school and all. Same thing happens when you start with a ferry on a river. A town begins, where there wasn't anything but wild country.

— from *Wagon Wheels: A Story of the National Road* by William A. Breyfogle

Jediah Hill and his family first appear in the records of Springfield Township in the 1820 federal census. They are a family of three: one man between the ages of twenty-six and forty-five, one woman between the ages of sixteen and twenty-six, and one girl under the age of ten.[19] The head of household's occupation was farming. At the time, Jediah was twenty-seven, Eliza twenty-four, and Rachel Maria four years old.

The family settled on 200 acres in Section 28 that was owned by Jediah's father and was bounded on the south by his brother Samuel's land in Section 27. The dawn of the 1820s must have been a time of great plans for the brothers. Local lore says Jediah built a cabin for his family, and then "set up a sawmill."[20]

Why Build a Sawmill?

Jediah Hill chose to start his mill at a time when competition among individual small business owners acted as a stimulus to innovation. This led to the Industrial Revolution, during which the nation saw an upsurge in inventive genius and ingenuity.[21]

In the early part of the nineteenth century, sawmills sprang up everywhere in America because of the abundance of timber, the relative simplicity of the saws, and the huge demand for boards and shingles.[22] It was customary to build a sawmill before a gristmill because the first provided building material when the miller was ready to expand his business to include the latter.[23]

Even though Jediah, who, based on his responses to the census taker, considered himself a farmer, he recognized that a sawmill business could provide him with a reliable supplemental source of income. His land had plenty of timber available — and he planned to clear it to plant crops, anyhow.

Mills were social centers of rural communities, and Jediah was already well connected in the community through his brother, Samuel. Many of their friends and relatives from New Jersey had also settled in the area, so he was hardly a stranger, even though he himself hadn't lived in Ohio for long.

But milling wasn't the kind of business one could start on a whim. Building the water wheel, gears, and delicately balancing the wooden equipment required technical knowledge.[24]

Though there is no evidence that Jediah had any prior experience with milling, he may very well have been apprenticed to a miller while a boy. East Amwell, New Jersey, was an established agricultural community where mills of all types were part of the landscape. He seemed to have had a close relationship to Squire Jacob Williamson who owned a mill in Clover Hill, New Jersey. On their trip east in 1838, Jediah's son-in-law, Henry Rogers, recorded in his travel journal that he and Jediah had paid two visits to Squire Williamson and spent the majority of that time looking over Williamson's mill.[25]

It is possible that Jediah had been apprenticed to or worked for Squire Williamson, and therefore had specialized knowledge that guided him in his own business venture. One thing has become evident: Jediah didn't undertake the mill project alone.

It would have been foolish to attempt to build and equip a mill without the advice and assistance of a competent millwright. When assessing the site for a mill, there was much to take into account. Each mill was custom designed to work with the terrain — even mills constructed by the same builder were never identical.

John Lane, Jediah's closest neighbor, is believed to have assisted in the siting and construction of other mills in Hamilton and Butler counties, and it is very likely that Jediah hired him to help with the construction of his mill. In the following decade, Henry Rogers and John Lane would work together to refine one of the most important inventions of the Industrial and Agricultural revolutions.

But first, Jediah Hill and John Lane combined their skills and talents to create a mill structure of such quality of workmanship that their legacy lasted for the entire life of the mill. The mill's longevity can be attributed to both the adaptability of the original structure and the ingenuity of each successive owner, all of whom made modifications to both take advantage of new technologies and meet the changing needs of the community.

Sidebar 2.1

The Lane Connection

The Lane family's history is closely interwoven with that of the Hills. Their futures and destinies were linked from the moment Aaron Van Doren sold 300 acres of land in Springfield Township Section 28 to Samuel and Jediah's father, Paul Hill. Paul later sold 200 acres of that land to Jediah.[26]

Aaron Van Doren's daughter, Sarah (1767–1800), married Aaron Lane (1763–1846) of New Jersey, and they brought their family to Hamilton County in around 1797 or 1798, and settled in what is now Springdale, Ohio.

Aaron and Sarah's two sons, John Lane and Aaron Van Doren Lane, inherited land in Section Twenty-eight from their grandfather.[27] The Lanes were neighbors of Samuel Hill's before Jediah arrived in Ohio in 1819.

John Lane married and settled on that land, and around 1813 opened a blacksmith shop (Figure 2.1), the ruins of which are still in evidence near the family home, built around 1840.[28]

Figure 2.1. This image, dated from the late 1860s, shows the John Lane home. The stone cottage blacksmith shop is to the left of the house. Both structures remain standing today, in states of decay. (Image courtesy of Pat Brown of Brown Studios Photography, Hamilton, Ohio.)

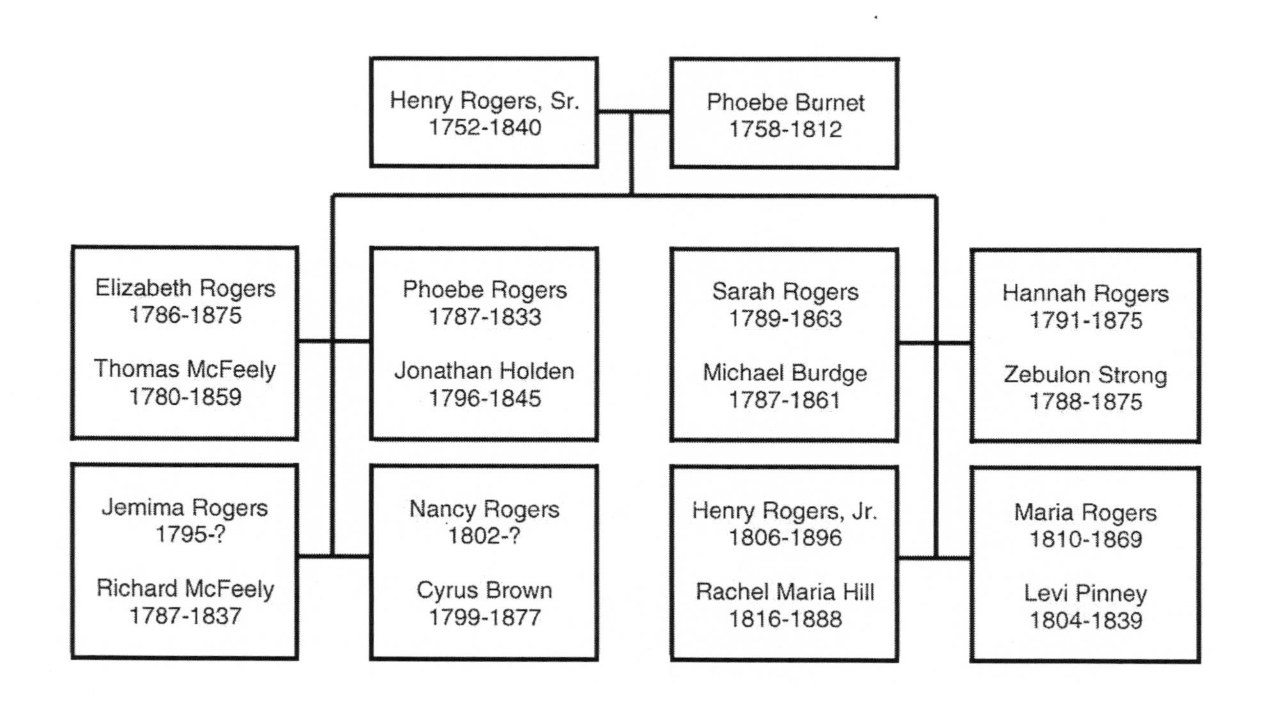
Henry Rogers, Sr.
1752-1840
Phoebe Burnet
1758-1812
Elizabeth Rogers
1786-1875
Thomas McFeely
1780-1859
Phoebe Rogers
1787-1833
Jonathan Holden
1796-1845
Sarah Rogers
1789-1863
Michael Burdge
1787-1861
Hannah Rogers
1791-1875
Zebulon Strong
1788-1875
Jemima Rogers
1795-?
Richard McFeely
1787-1837
Nancy Rogers
1802-?
Cyrus Brown
1799-1877
Henry Rogers, Jr.
1806-1896
Rachel Maria Hill
1816-1888
Maria Rogers
1810-1869
Levi Pinney
1804-1839

Henry Rogers, Sr. — Revolution, Relocation, and Re-Interment

3

Henry Rogers, Jr., began his long association with Jediah Hill as his hired hand, and later became his son-in-law and business partner. Jediah may have been considered an early pioneer to the area, but Henry had come to Ohio as an infant, more than a decade before Jediah brought his family west.

His father, Henry Rogers, Sr., was born in Middlesex County, New Jersey, on December 21, 1752, and was a weaver by trade.[29] Besides a brief biographical mention in *History of Hamilton County*, most of what is known about the elder Rogers is found in his Revolutionary War pension application, in rosters of Revolutionary War soldiers, and in National Society of the Daughters of the American Revolution application forms.

I learned that he enlisted in the First New Jersey Regiment, known as the Jersey Blues, in the autumn of 1775, when he was twenty-two years old.[30] Eight companies of the First New Jersey were raised from men residing in Essex, Middlesex, Morris, Somerset, Monmouth, and Bergen counties. His commanding officer was Colonel William Alexander, also known as Lord Stirling. Once mustered, Henry's unit set to work subduing and capturing Tories on Long Island.[31]

At the Battle of Long Island in August, 1776, Stirling's troops were outnumbered twenty-five to one, and he was eventually captured, but not before he repelled the British forces until the main body of his troops could escape to defensive positions at Brooklyn Heights. After the battle, Stirling was praised for his bravery by both George Washington and the British.

When Henry's enlistment expired in November, 1776, he returned to civilian life. His skill as a weaver made him more valuable to the war effort at the loom and off the field of battle.[32] He married Phoebe Burnet in 1787, when he was thirty-five and she twenty-nine, and together they had seven daughters and three sons, of which Henry Jr., was the youngest. His older brothers, Amos and Samuel, died in infancy.[33]

The Rogers family migrated from either Pennsylvania or New Jersey in 1806,[34] and paused along their way in Greensburg, Fayette County, Pennsylvania, where Henry was born on May 31 of that year, joining sisters Elizabeth, Sarah, Hannah, Phoebe, Jemima, and Nancy. It appears the Rogers family may have lingered in Fayette County for several months, as Elizabeth, the eldest daughter, wed Thomas McFeely there on January 20, 1807.[35]

The Rogers family settled in Mill Creek Township in Hamilton County, Ohio,[36] where their daughter and new son-in-law also appear to have arrived sometime in 1807. Their first grandchild, James McFeely, was born there on November 9, 1807.

Maria, the youngest daughter of Henry Sr. and Phoebe Burnet Rogers, was born in Ohio in 1810. Phoebe died in 1812 at the age of fifty-four.[37]

There are no records to give us a glimpse of Henry Jr.'s childhood, but local lore says he left home at seventeen years of age, "a poor boy, to fight the battle of life alone."[38] He is supposed to have learned the weaver's trade from his father,[39] and also was reportedly apprenticed to a cabinetmaker, but left the apprenticeship because he did not care for the trade.[40] Apprenticeships varied in length, but often the end was determined by the apprentice's twenty-first birthday. If Henry was apprenticed at seventeen, his agreement may have been for four years. It's impossible to determine exactly if, where, or how long Henry attended school, but whatever formal schooling he received was surely supplemented at home, as he developed into an eloquent writer and a keen observer of the world around him. In his adult life, he also demonstrated a talent for carpentry and mechanics.

It is evident that Henry's family valued education, as all but one of his sisters who were alive for the 1850 federal census were listed as literate. Nancy Rogers Brown and her husband, Cyrus, indicated they were unable to read and write in the 1850 census, but in the 1860 census they were identified as literate, so perhaps the 1850 census taker had erred.

I need to direct our focus back to the Hill family for a moment and mention that the 1830 federal census of Springfield Township shows Jediah Hill (written Jedediah on this document) surrounded by names that are now familiar: his widowed sister-in-law, Mary Hill, her brother, John Woolverton, fellow pioneer Aaron Lane and his son Aaron Van Doren Lane, and physicians Thomas and Richard McFeely (written McPheely).[41]

Thomas McFeely had married Henry's eldest sister Elizabeth, and his brother Richard had married the fifth Rogers sister, Jemima. It is likely that either or both Elizabeth or Jemima introduced Henry to Jediah Hill, their neighbor who was in need of a hired hand.

In time, all eight of the Rogers children married, and at least seven of the resulting sons were named for their grandfather, Henry. Elizabeth and Thomas McFeely had sons named John H. and Joseph H.; we may assume the middle names of both were Henry.[42] Phoebe married Jonathan Holden, and they named their sons Henry R. Holden and James Henry Holden.[43] Sarah married Michael Burdge, and they named one of their sons Henry Rogers Burdge.[44] Hannah married Zebulon Strong. They named one of their sons Henry Rogers Strong, and one of their daughters Phoebe Burnet Strong. It should also be noted that they named another of their sons William Cary Strong, after their friend, the noted abolitionist and educator.[45] William Cary founded the town of College

Hill, home of Cary's Academy, Farmers College, and Lane Seminary, on his own land.[46] Nancy married Cyrus Brown. There is no record of issue from this union.[47] Maria, the youngest daughter, married Levi Pinney, son of one of the founders of Worthington, Ohio. Levi and Maria named their second son Henry Pinney.[48]

The western migration that had brought the Rogers family to Ohio was bound to continue. Over time, the children scattered. Henry's sisters Elizabeth, Jemima, and Phoebe and their families moved farther west, to Indiana and Illinois. Sarah and Maria and their families resided in Columbus, while Hannah and her family lived only a few miles away in College Hill. Nancy and her husband, Cyrus, remained close by, on their farm in Springfield Township, and later in College Hill, before moving west to Montgomery, Indiana.[49]

Henry Jr. had found his place in the world on the banks of Mill Creek, where he would start his working life alongside Jediah Hill, the farmer, miller, and entrepreneur, who would come to appreciate having a skilled carpenter in his employ.

In September, 1826, when he was seventy-four years old, Henry Sr. applied for a Revolutionary War pension, and was awarded an allowance of $8.00 per month. He was also awarded a semi-annual arrears payment of $40.25, to be paid from 1826 to March 1829.[50]

As Henry Sr.'s children built their own lives and fortunes, he seemed to slip into obscurity, first in life, and then in death. The 1810 federal census for Ohio, the only time during which we can assume Henry Sr. was head of his own household, is lost, except for Washington County. His name does not appear in the 1820 or 1830 census, but it is possible that, as a widower, he was living in one of his daughters' homes. He does not appear in the 1840 census, and may have died before the census taker arrived that year.

Dead and Forgotten . . . for a While

When Henry Rogers, Sr., died in Hamilton County, Ohio, on July 17, 1840, at the age of eighty-seven, he was buried in the Quaker Cemetery on Drehman Road in Cumminsville, also known as the Roll Cemetery.[51]

While preparing for a research visit to the area in 2014, I looked up the location of the Roll Cemetery, hoping to pay my respects at Henry's grave and find him honored with a bronze marker placed by the Daughters of the American Revolution, and maybe even an American flag reflecting his service. But a Google search for the Roll Cemetery led me to two articles that appeared in the *Cincinnati Post* in the summer of 1923. Eighty-three years after Henry's death, the tiny cemetery where he was buried was embroiled in a scandal.

> **"Asks $25,000 for 'Desecration' of Quaker Cemetery.**
>
> Suit for $25,000 damages, for alleged desecration of the old Quaker cemetery, was filed in Common Pleas court Wednesday by James S. Myers, former magistrate...
>
> Myers alleges that his grandfather, Simon Myers, was buried there in 1835, and that his grandmother,

Elizabeth Myers, was buried there in 1838. He said that the cemetery, which was purchased in 1834, was to be used for burial purposes only.

Bones Piled in Building

He alleges that the trustees of Millcreek Township sold the cemetery on January 12, 1923, to [Charles E.] Miller, who removed the headstones and dug up the human bones buried there and "placed them in an indiscriminate heap in a shed on the premises, so that the plaintiff cannot distinguish the remains of his grandparents from the other human bones piled there."

A follow-up article by *Post* columnist Alfred Segal, dated August 1, 1923, revealed a gruesome discovery.

> "Edward Roll, a citizen of Cincinnati, died 103 years ago and was buried in the Quaker cemetery at Drehman ave and West Fork Creek. He was 48 years old.
>
> Yesterday the Post's photographer. . . uncovered [Roll's] tombstone and read the pledge of remembrance [etched on the stone]. And a few feet away lay a skull and a jawbone. It may be presumed that these are the bones of Edward Roll, for they are like all the other bones that lately have been dug from the cemetery of the Quakers, and no one can tell which was the rich man's bones and which the poor one's; which the wise one's and which the fool's."[52]

Henry Sr.'s remains were probably among those removed from their graves and stored in a shed by the Miller Sand Company in the summer of 1923. Surely the bones had been reburied to satisfy the injured parties — but where? I consulted Ancestry.com and the Find-a-Grave Index database, which confirmed that Henry Rogers, Sr.'s, remains were removed from Roll Family Cemetery and now rest in Wesleyan Cemetery in Cumminsville, which is in Mill Creek Township. No plot or grave number was indicated, and Henry's name was not included on the map showing veterans' graves.[53]

The Cincinnati Chapter Sons of the American Revolution have since undertaken to honor and memorialize the eighteen Revolutionary patriots that lay buried in Wesleyan Cemetery, and on October 18, 2015, dedicated a granite monument there (Figure 3.1).

James D. Schaeffer, President of the Ohio Society of the Sons of the American Revolution, spoke at the ceremony.

> "Through this monument placed here today, we satisfy a long overdue obligation to pay homage to these patriots. We seek to do something that is permanent, something that will tell the stories of significant and meaningful lives to future generations. May we always honor the sacrifices made to preserve our freedom by all patriots, past and present."[54]

Henry Rogers, Jr.'s, 1838 travel journal demonstrates that both acknowledgement of one's military service and an

Figure 3.1. Monument to Revolutionary War soldiers at Wesleyan Cemetery, Cumminsville, Ohio.

Figure 3.2. The Revolutionary Soldiers' Cemetery, East Vincent Township, Pennsylvania.

appropriate burial place were very important to him. In his entry dated September 8, Henry makes special note of a visit to a Revolutionary soldiers' cemetery in East Vincent Township, Pennsylvania (Figure 3.2). He mentioned an obelisk honoring twenty-two patriots who had died of a fever in the winter of 1777 and wrote:

> "Those 22 true Sons of Columbia were lodged in the Dutch Reform Church in sight of the Valley Forge, which was then occupied as a hospital for the American Army. There they were, far from their homes, uncomforted by friends and no smooth soft

hand of affection to console them in their dying moments." [55]

He also noted that the Chester County, Pennsylvania, Battalion of Volunteers had erected the monument "so that the present and future generations yet unborn might see where and under what circumstances those true and patriotic Sons of Liberty and Independence perished." [56] Knowing how Henry Jr. felt about honoring these departed veterans, it is fitting that his own father is now properly acknowledged and honored for his service to his country.

Sidebar 3.1

Overland Travel

In 1790, the center of population in the young United States was in eastern Maryland; in 1800, it was just north of Washington, D.C.; by 1820, it had moved to what is now West Virginia; in 1860, it was far into Ohio. It had moved at a rate of approximately five miles a year to the heart of the American continent.[57]

In the 1790s, Pennsylvania and other states began massive road-building campaigns to improve trade and open new markets in hard-to-reach areas. The country's infrastructure and roadways would see another developmental surge following The War of 1812.[58]

In 1800, it took about three weeks to travel between Philadelphia and Pittsburgh if the weather was favorable. A paved road was completed by 1820, but in 1806, Henry's family would have followed various unpaved Native American paths and war routes.[59] A wagon trail stretched from Philadelphia west over the Appalachian Mountains to Fort Pitt, now the city of Pittsburgh.

There is no way to know exactly where the Rogers family paused in Greensburg, Pennsylvania so that Henry Jr. could be born, but a tiny settlement, known as Newtown, grew around an inn. Today, the location of that inn marks the center of Greensburg's business district, at the intersection of Pittsburgh and Main streets.[60]

By the time Jediah Hill and his family came west in 1819, the National Road had extended from Cumberland, Maryland, as far as Wheeling, Virginia. The groundbreaking for the National Road in Ohio took place on July 4, 1825, and the Hill and Rogers families would use the completed eastern section of the Road for much of their 1838 journey.[61]

Sidebar 3.2

Lord Stirling

Colonel William Alexander, Henry Sr.'s commanding officer in the First New Jersey Regiment (Figure 3.3), was a bright and wealthy man who had inherited a large fortune from his father and claimed the disputed title Lord Stirling.

He lived a socially prominent life, much as befitted a Scottish lord.

Figure 3.3. William Alexander, Lord Stirling, commanded the First New Jersey Militia.

In 1747, he married Sarah Livingston, the sister of William Livingston, who served as Governor of New Jersey during the Revolutionary War. He partied with George Washington and was awarded a Royal Society of Arts award in 1767 for establishing winemaking in the colonies.

When the war began, Alexander was made a colonel in the New Jersey Militia. He outfitted his men at his own expense and was always willing to spend his private fortune in support of the revolutionary cause.[62]

4 Henry Rogers, Jr., and the Obed Hussey Reaper

What does modern machinery mean to the farmer? In 1830, a farmer in a working day of ten hours produced three bushels of wheat. In 1920, a farmer in a working day of ten hours produced sixty bushels of wheat.

— C. B. Benson, author of "A Northern California Herd."

In 1800, eighty-three percent of the labor force of the young United States worked in agriculture.[63] The tools of farming had stayed essentially the same for ten centuries. It's hard to believe that the farmers of George Washington's day (Figure 4.1) had no better tools than the farmers of Julius Caesar's day.[64] In rural communities, it was common for craftsmen, and even doctors, lawyers, and ministers, to farm part-time as a means of survival.[65]

Figure 4.1. Farming methods, like those depicted in this print of a wheat harvest at George Washington's Mt. Vernon, remained essentially the same for ten centuries.

By 1830, the early effects of the Industrial Revolution that had begun at the end of the eighteenth century were evident. Workers were needed in factories, and the percentage of the population working in agriculture dropped to seventy-one percent.[66] The size of the average New England farm was between 70 and 120 acres,[67] and it took between 250 and 300 hours of labor to produce 100 bushels of wheat, with an average yield of five bushels per acre.[68]

Growing wheat in pre-industrialized times was a labor-intensive process. Someone had to use a horse or oxen and a walking plow to break the ground. Then they had to sow the seed by hand, hoe weeds and cultivate by hand, cut by hand with a sickle, and use flails[69] to shake the grain loose from the stalks (figures 4.2 and 4.3).

When work-saving farm implements became available early in the nineteenth century, the way people worked and lived in America saw rapid change. One of the first innovations to hit the market during this time was the Obed Hussey Reaping Machine.

> "It was in the middle of June in 1835. The farmers and mechanics that had gathered at Jedediah (often given as Jediah) [sic-Jediah was correct] Hill's farm north of Mt. Healthy, Ohio, were in high spirits. Inventor Obed Hussey, blacksmith and farmer John Lane, farmer and mill owner Hill, Hill's son-in-law

Figure 4.2. Workers cut grain by hand before the invention of the reaping machine.

Figure 4.3. Thrashing (threshing in the modern spelling) is the process by which grain is removed from the stalk. The first thrashing machine was invented in 1786 by Scottish mechanical engineer Andrew Meikle. Mechanization of this process took much of the drudgery out of farm work and led the way for more innovations in the nineteenth century.

Henry Rogers, farmer Algernon Sydney Foster, and others hitched the newly minted reaping machine to a team of horses. Surrounded by ruddy-cheeked men and cheerful boys, including Lane's sons, the reaper entered the barley field (Figure 4.4).

Soon, the invention with its iron teeth was mowing the crop and leaving the stubble behind in its wake. Hussey urged the draft horses to a pace faster than that to which they had become accustomed. Wide eyed and snorting their surprise, they clipped off a rapid stride. The men nodded to one another as the snappier speed cut the stalks perfectly.

Were these men aware that they had just made world agricultural history? . . . that their names would be recorded forever? . . . that the destiny of the Earth's civilization was in their hands? Probably not. We recently walked across a covered bridge and stood at the edge of the former barley field. Ironically, the farm that had been cleared of trees by arduous labor when the United States Constitution was drafted has become a forest again. Not far from the famous barley field, ranch houses with clipped green lawns border the tangled thickets draped with wild grapevines and Virginia creeper. The homes face across the road to the site where Hill's mill stood and where Rogers, with consummate skill, finished the reaper under Hussey's supervision. Despite the presence of tall trees and undergrowth, we could picture the momentous day 179 years ago. With sunlight and shadows of clouds dappling the stones of Mill Creek and robins singing their morning songs, we silently contemplated the fact that our feet were planted where the feet of some of the greatest inventors of the Industrial Era had walked. For it was in that barley field that the first successful reaper, a machine that would initiate an agricultural revolution, was tested!"[70]

Figure 4.4. This print of the Lane house and blacksmith shop was featured on the cover of *Deering's Farm Journal* in May, 1838. The first test of the Hussey reaper, which took place in Jediah Hill's barley field, is depicted at the upper right.

I'd been researching and writing about my family for over two decades. How had I never heard this story before? This event should be a celebrated part of Mount Healthy's history! How thrilling that Rhode and Hite heralded Obed Hussey, John Lane, and Henry Rogers as "some of the greatest inventors of the Industrial Age."

A commemorative pamphlet entitled *The Story of New Burlington, 1816–1922* makes brief mention of the reaper.

> "The little hamlet has given its portion of history to the country. . . . Mr. John Lane built the little stone cottage which stands on the southeast corner of Mill Road and Hamilton Pike (Figure 4.5). Here he kept a blacksmith shop where the first reaper was built." [71]

Clark Lane, inventor and businessman and one of John Lane's sons, noted in his unpublished autobiography:

> "My Father's Book of 1835 now before me, though crude of keeping does witness that early in that year he and my present Octogenarian friend and neighbor Henry Rogers constructed and put into practical and successful operation what I am quite sure was the first Reaping Machine ever built upon, and that cut grain within limits of the Northwest Territory." [72]

How had this important moment in American history — one that had literally happened in Jediah and Henry's front yard — been downplayed and nearly forgotten?

Though I'd never heard of Obed Hussey's invention, I *had* heard of a McCormick Reaper. How did the McCormick machine eclipse Obed Hussey's — both in the market and in the annals of history?

During the winter of 1832–1833, Obed Hussey, a Maine-born Quaker, had moved from Baltimore to Cincinnati to work on the reaper that would bring him fame during the harvest of 1833. The April 1834 edition of *Mechanics Magazine*'s contains an illustration of "Hussey's Grain Cutter" (Figure 4.6). A notice in that same edition of *Mechanics Magazine* stated:

Figure 4.5. This plate from Follett L Greeno's biography of Obed Hussey shows his portrait and the Lane blacksmith cottage, where parts for the prototype were forged.

"This may certify, that we, the undersigned, members of the Agricultural Society of Hamilton county, state of Ohio, at the request of Mr. Obed Hussey, attended an exhibition of a machine for cutting grain by horse power, invented by him. The experiment was performed at Carthage, in this county, about the first of July [1833] before a large company of spectators . . . who appeared to be united in the expression that it was a valuable improvement in agriculture."[73]

The Agricultural Society went on to point out that ". . . several impediments occurred during the exhibition by the breaking of some weak parts."[74]

Hussey had competition, as Robert and Cyrus McCormick had produced another version of the reaper at about the same time. Although McCormick's first patent was dated 1834, when he applied for his extension in 1848, he alleged that his reaper was invented prior to Hussey's in 1831, and three years before the date of his own first patent. The Hussey machine's metal cutter bar, with teeth that moved back and forth, was what made Hussey's invention unique — and superior to McCormick's.

McCormick presented evidence to support his claim of priority, but it was deemed inadmissible by the Board of Review at the United States Patent and Trademark Office. The Patent and Trademark Office refused to go on with the examination either as to priority or validity of invention without notifying Hussey, because McCormick's claim called Hussey's patent into question.[75]

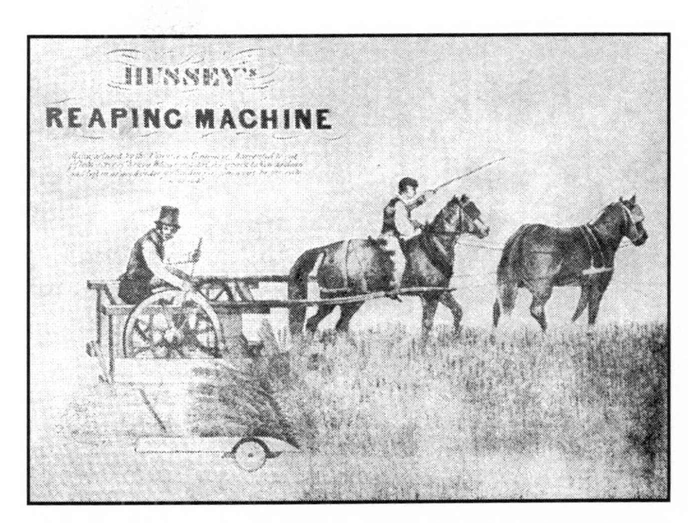

Figure 4.6. Undated advertisement for the Obed Hussey reaper.

During the ensuing patent wars between McCormick and Hussey, scores of customers wrote extolling Hussey's reaper as the far superior product. Through all the attacks, Hussey never sought to take away from McCormick's reaper, and merely wanted his own invention put to use by the farmers for whom it was created. McCormick's aggressive tactics eventually drove Hussey out of business, and McCormick was able to acquire the rights to Hussey's superior cutter-bar mechanism around 1858. Only then did the McCormick Reaper experience mainstream success and acceptance.[76]

After Obed Hussey died in a train accident in 1861, the United States Patent and Trademark Office issued a ruling, which determined that the money made from reapers was due in large part to Obed Hussey's invention. The acting commissioner of patents declared that Hussey's improvements were the foundation of the machine's success. The ruling further declared that the heirs of Obed Hussey should be compensated for Hussey's hard work and innovation by others who made money from sales of the reaper.[77]

Hussey's death did not quell the debate over who deserved credit for inventing the first reaping machine, as witnessed in this letter written by Clark Lane, son of John Lane:

> Editor Democrat:
>
> To "Inquirer who from Oxford, Ohio asks the DEMOCRAT for facts as to 'the origin of the Reaper and Mower'" please say that on the Cincinnati and Hamilton pike midway between the two cities, may yet be seen the buildings in which the first reaper was made; i.e. it was the first reaping machine made upon terra firma of Northwest territory, and that ever cut grain north, west, or southwest of the Ohio river in manner and condition to assure the farmer that the time had come when his grain could be economically harvested with machinery.
>
> Prior to July 1835, in the stone building not twenty (20) paces from where I write, my father, John Lane, forged and furnished the iron and steel parts for said reaper and Henry H. Rogers, at his place of business not five hundred paces distant finished and fitted up both iron and wood work to a finish and working condition of the same.
>
> The work was done for Jefferson and Algernon Foster, more or less under the personal superintendence of Obed Hussey, who furnished patterns, drawings, and some of the castings used.
>
> In the first ripening grain of that year on the farm of Jedediah Hill, and now the home of Mr. Rogers, the Obed Hussey machine made its first cuts and proved its great superiority over all former efforts in this class of invention and manufacture.
>
> March 10, 1890"[78]

The reaper was one of the most important inventions of the Agricultural Revolution in the first half of the nineteenth century. By 1840, the percentage of workers engaged in agriculture had dropped to sixty-three percent, and the number continued to fall even as the number of farms increased. New states were admitted to the Union, and for farmers all over the nation, the reaper and other factory-made agricultural machinery decreased the need for human labor and encouraged commercial farming on a larger scale.[79]

Henry helped refine the Obed Hussey reaper and get it ready for its first trial in 1835. If he and Jediah weren't already thinking ahead toward converting the mill to grind flour, surely this experience pushed them in that direction.

It might have seemed like a sure bet — farmers who purchased Hussey reapers would clear more land and be able to harvest more grain more easily than ever before in the history of agriculture. There would then be greater demand for the services of gristmills. But it seems nothing goes as smoothly as it ought.

From 1839 to around 1847, local contractors in Hamilton, Champaign, Jefferson, and Stark counties in Ohio manufactured the Hussey reaper. McCormick's machines were manufactured at Cincinnati beginning around 1845. During the next few years some of the machines were produced by subcontractors within the state, and quality suffered. By 1851, relatively few of both the McCormick and Hussey machines were in use.[80]

But back in the fall of 1835, there was every reason for optimism. Jediah and Henry set their sights on a huge structural addition and other capital improvements to the mill, so they would be ready for the anticipated boom in agricultural production. Chief among their concerns was how to harness enough hydropower to enable them to run the mill for more than three or four hours a day, which Steve Hagaman estimates was its capacity during rainy season, and how to keep running at all during times of drought.

There were plenty of mills doing business along the National Road, which extended from Cumberland, Maryland, to as far as Columbus, Ohio, by 1834. Jediah still had plenty of family and friends in New Jersey and Pennsylvania. It was time for a trip back east. They could observe and learn about other mills along the way.

Sidebar 4.1

John Hall's Sickle

John Hall was born in Uniontown, Pennsylvania, in 1794 and came to Hamilton County with his family in 1811. They settled in what is now Glendale, where John used this sickle (Figure 4.7) to harvest wheat on the family farm.

John Hall is an example of a man, a contemporary of Jediah Hill's, who was a direct beneficiary of the fruits of the labors of the inventors and innovators of the early nineteenth century. It is possible the sickle survived to this day because John Hall hung it up in favor of some labor saving device. Like, perhaps, a reaping machine.

This link to the past is even more delightful when you know that John's granddaughter, Laura Carroll, married Jay Rogers, a grandson of Henry Rogers.

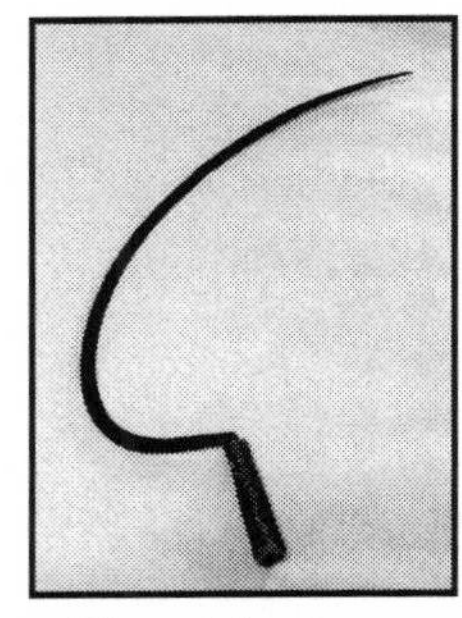

Figure 4.7. John Hall's sickle. Image courtesy of Glendale Historic Preservation.

SIDEBAR 4.2

The Lincoln Connection

Obed Hussey and his reaper were not Cyrus McCormick's only targets in his battle to claim the patent on the machine.

John H. Manny had developed a similar reaper in Wisconsin, and moved his factory to Rockford, Illinois, effectively encroaching on McCormick's home turf. McCormick sued for infringement and damages of four hundred thousand dollars, and hired nationally known lawyers from New York and Washington. The Manny Company hired two of the country's leading patent attorneys, plus Edwin M. Stanton of Pittsburgh.

"The Reaper Case" attracted national attention and would set precedent for how patents would be protected in the growing farm implement industry.

Because the case was to be tried in federal court in Chicago, the Manny Company also retained a local lawyer who would have the trust of the federal judge. They chose Abraham Lincoln, who had recently won a seat in the Illinois House of Representatives and was on the rise as an attorney and politician, but was still considered by some to be awkward and countrified. As the Illinois lawyer on the team, Lincoln was to give the closing argument.

The case was set for trial in 1855, and in the months preceding Lincoln toured the Manny factory to become familiar with the reaper, and gave his case careful preparation.[81]

Just weeks before the trial was to commence, its venue was moved from Chicago to Cincinnati. The legal team no longer had any use for Lincoln, but failed to apprise him of the changes. When Lincoln arrived in Cincinnati for the trial, lead attorney Edwin M. Stanton, who less than ten years hence would be Secretary of War in President Lincoln's Cabinet, snubbed him, ignoring the lengthy brief he had prepared. The Manny team of lawyers told Lincoln he would have no role in the case. He was not invited to join the other lawyers for meals at the hotel or to take part in deliberations.

Despite this treatment, Lincoln attended all court sessions as a spectator. After the trial, which was decided in favor of the Manny Company, Lincoln was sent a check as payment for his participation in the case. He initially refused it on the grounds that he had no right to any fee beyond the original five hundred dollar retainer. Eventually, he was persuaded to take the payment for his time spent in preparation.[82]

5 Cross-Country Journey — 1838

As it is my intention to mention all interesting subjects and things that come under my observation, I will endeavor to describe our course.

— Henry Rogers, travel journal, August–October, 1838

In the eighteenth century, eastern Pennsylvania was considered the breadbasket of the colonies, and then of the United States, with great wheat fields supporting flourishing flour milling industries, but by the 1840s, that designation had shifted west to the Ohio Valley.[83]

In 1810, fifteen percent of the American population lived west of the Appalachians, which by then including the newest state, Ohio. Within a decade, six more western states would be added to the Union. Cincinnati, the Queen City of the West, had replaced Lexington as the region's cultural and commercial epicenter, with a population that burgeoned from 24,831 in 1830 to 46,338 ten years later.[84] By 1840, more than a third of all Americans lived in this First West[85] (Figure 5.1).

Figure 5.1. 1848 Daguerrotype of Cincinnati, made ten years after the Hills and Rogers went on their cross-country journey, shows a thriving city worthy of the name Queen City of the West.

Most of the original Europeans who settled Ohio were farmers who raised wheat, corn, and other grain crops. By the middle of the 1800s, agriculture had flourished to the extent that Ohio produced more corn than any other state, and ranked second in wheat production.[86]

It was after the publication of *Fips, Bots, Doggeries, and More* that I learned of Henry's involvement in the Obed Hussey Reaper's development. Jediah and Henry's knowledge of the machine's capabilities and its potential impact on agricultural methods was probably the catalyst that led them to consider expanding their business into a saw-and-grinding establishment. I continued to search for evidence that would prove exactly when that expansion took place.

Renovating the existing mill to grind flour would require a massive, multi-story addition to accommodate millstones, elevators, chutes, and machines to clean the wheat and sift the flour.

Once the plan to seek out and observe working mills was under consideration, I wonder if Jediah's wife Eliza might have mentioned how much she missed the home folks they'd left behind in New Jersey. I'm going to put words in her mouth for a moment because, if I'd left home as a young wife and mother to help my husband build his fortunes on the frontier, I might feel I had the right to mention how much I'd like to see my mother and my brothers and sisters again. "Think how much the country's grown up since we came west, Jediah. Maria was too young to remember the journey, and Henry was a mere babe in arms when his folks came to Ohio. Maybe we should all go home for a spell."

There had never been a better time for Ohioans contemplating overland travel to go back east, but even so, the family's planned route from Mount Pleasant to New York City and back would take nearly three months. They planned to leave in mid-August, and as they wouldn't be returning until late November, would have needed to pack both summer and winter clothing. They would have brought tools to repair and maintain the wagon, and provisions for themselves and their horses, in case there were no accommodations available at the end of a day's travel.

The journey got underway on August 18, 1838. The planned route first took them north, to visit family in Warren and Miami counties. They then traveled east along local roads until they came to the National Road at what is now West Jefferson, Ohio. Their course of travel continued on the National Road through Ohio, what is now West Virginia, Pennsylvania, and Maryland (Figure 5.2). When the National

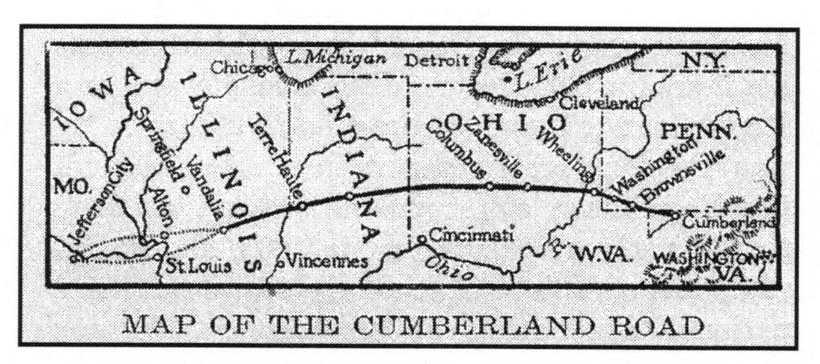

Figure 5.2. The National Road extended as far as Columbus, Ohio by 1834, and made travel to and from the new states west of the Alleghenies easier than ever before. Image courtesy of Eric Fisher.

Road terminated at Cumberland, Maryland, they continued east on local roads through Hagerstown, then turned northeast through Leitersburg, Maryland, and into Pennsylvania, where their route took them north through Gettysburg and other small towns until they came on what is now Route 30, the Lincoln Highway. They continued east through the old cities of York and Lancaster and then passed through the Conestoga Valley. They bypassed Philadelphia to the south, and arrived at a relative's home in Trenton, New Jersey on September 22. The remainder of their time was spent visiting with relatives in Trenton and Philadelphia, as well as sightseeing, shopping, and visiting mills.

Three years had elapsed between June, 1835, when the reaper proved its mettle in Jediah's barley field, and August, 1838, when the family embarked on their journey east. This indicates the decision to undertake the cross-country journey was not to be made lightly.

It is possible that their enthusiasm for expanding the mill could have been tempered by the land boom and bust and the recession that led to the Panic of 1837.

Some economists and historians blame President Andrew Jackson and his Specie Circular Act for the country's economic woes, while others absolve him of blame. As always, more than one factor contributed to the country's economic downturn. The United States experienced an influx of silver coin from Mexico in the mid-1830s, which was circulated as legal tender and increased the overall specie supply, and that in turn led to increased credit, rising prices, and a business boom. The total money supply in the United States rose by eighty-four percent from 1833 to 1837, and could have set up conditions for a boom-and-bust cycle.[87]

Paper money issued by the government was declared irredeemable for gold or silver [specie] during 1837, which devalued paper money and drove the symptoms of the Panic. Limited specie payments resumed in the summer of 1838, which led to a large outflow of coin, and the exchange of specie for paper money was again suspended from 1839 to 1842.[88]

Jediah and Henry may have hesitated about making the expensive trip, to say nothing of contemplating the costly improvements to the mill, as they followed the ups and downs of the recession. Perhaps they ultimately decided to live by Benjamin Franklin's proverb, "Nothing ventured, nothing gained."

Sidebar 5.1

The National Road

The National Road, America's first federally funded highway, stretched westward from Cumberland, Maryland, to its eventual terminus at Vandalia, Illinois. Construction of the road, started in 1811, had halted during The War of 1812, and resumed in 1818. The groundbreaking for the Ohio portion of the road took place July 4, 1825. Engineers and construction crews pushed westward across the middle of the state, and the Road was open as far as Columbus by 1834 (figures 5.3 and 5.4). The Hills and the Rogers were able to travel along a recently constructed, fairly well-maintained road for a major portion of their journey.

Figure 5.3. Henry mentioned stopping at tollhouses like this one that stood near Frostburg, Maryland. Tollhouses were spaced about ten miles apart. Only three remain standing along the old National Road.

Figure 5.4. Curved, or S bridges, were common along the National Road, where survey teams marked the locations for bridges at right angles across creeks. Engineers built them ahead of the road crews, who later curved the road to meet the bridges when necessary.

Sidebar 5.2

The Specie Circular Act

Article I, Section 8 of the United States Constitution reads:

> "The Congress shall have power to lay and collect taxes, duties, imposts, and excises, to pay the debts and provide for the common defense and general welfare of the United States . . . To coin money, regulate the value thereof, and of foreign coin, and fix the standard of weights and measures."

The framers of the Constitution established a specie monetary system, based on coined money, to control inflation and avoid the financial problems experienced by the colonies during the Revolution, when the government had met its need for revenue by issuing irredeemable paper money. Under the specie system, bank notes and government-issued paper money could be redeemed for gold or silver coin at the issuing bank or the US Treasury.

President Andrew Jackson despised paper money and opposed the Second Bank of the United States, which controlled the flow of gold and silver and influenced the value of the state-run banks' paper money. Jackson thought the Second Bank was corrupt for contributing to the political campaigns of officials who then helped it maintain its power, and he planned to eliminate the Second Bank when its charter expired in 1836.[89]

To further cripple the Bank, Jackson issued the Specie Circular Act of 1836 by executive order. The Act required gold or silver as payment for the purchase of government lands. This Act was intended to curb inflation and protect settlers using devalued paper money to purchase farmland, and has been held responsible for the end of the land boom, the recession, and the Panic of 1837.[90]

In response to the Specie Circular Act, the Second Bank called back specie deposited in state-run banks, and this left the state banks with insufficient gold and silver to back their notes. The state banks called in loans made to settlers and farmers, who, unable to pay their debts, lost their homes and property. The Second Bank of the United States lost its charter in 1836, and, later that year, Nicholas Biddle, the director of the Second Bank, secured a charter from the Pennsylvania legislature to operate as a state bank. Though this United States Bank of Pennsylvania was influential for a few years, it lost money and went bankrupt in 1841.[91]

6 The Jediah Hill Era — Setting the Record Straight

Whether wittingly or unwittingly, facts may be distorted in historical accounts. Mistakes are perpetuated and accepted as fact. In conducting my own research, I noted a number of errors that appear in my family's biographical sketches. These mistakes may seem as though they make no difference now, but nonetheless, I believe in setting the record straight whenever possible.

It saddens me when the names of the women are incorrectly reported. The women I mention in this book didn't hold jobs that distinguished them in the community, they didn't leave journals or wills, and there are few surviving photographs. Their mention in historical records is seldom more than a report of to whom they were married. The least I can do is make sure their names and their spouses are sorted properly.

The earliest narrative about the Hill and Rogers family appears in *History of Hamilton County, Ohio* . . . by Henry A. and Kate B. Ford, and contains no known factual errors. The book was published in 1881, and the biographical information was gathered from primary sources.

Hover's *Memoirs of the Miami Valley* was published in 1919, thirty-eight years after *History of Hamilton County, Ohio.* The information on the Hill and Rogers families that appears in this book was likely provided by Wilson Rogers or his children, and is still quite accurate, though it incorrectly reports that Henry Rogers received his education in Pennsylvania before coming to Ohio as a young man.[92]

A number of discrepancies appear between these two accounts and *The Story of New Burlington, 1816–1922.* The errors perpetuated in later historical accounts seem to stem from this document:

> "Mr. Hill and his wife, Rachel Maria [wrong — his wife was Eliza Hendrickson][93] came to Springfield Township from New Jersey. He came to settle in the wilderness where the mill now stands. At first the settlers built a log cabin in which to live. Later Mr. Hill built the saw mill. Here he turned the logs of the forest into lumber from which he built his barn. They could live in the log house, but they could not shelter their stock nor store their grain. Sometime prior to 1843 the house was built. Later Mr. Hill also had a grist mill, where the grain of the neighborhood could be converted into flour.

With the passing of the years Mr. Hill's daughter, Maria, grew into womanhood. About this time, Mr. Henry Rogers, the son of a Revolutionary Soldier, came from the East [wrong — Henry had lived in nearby Mill Creek Township since he was an infant. His family had come to the area in 1806–1807] in search of employment. He engaged to work for Mr. Hill. Later he married the only child, Maria. [Henry refers to his wife as Maria in his journal, her name appears as Rachel M. Rogers or Rachel Maria Rogers on legal documents, and her headstone reads Rachel M. Rogers. It appears that though she was named Rachel after both her grandmothers, her family called her by her middle name. This likely added to the confusion about mother and daughter.][94]

The young people settled for life at the Hill home, where their only son, Wilson, was born and reared. Here he, in like manner brought up his family. The substantial old homestead has sheltered four generations."

A 1988 article by local historian Carolyn Kettell states that Jediah Hill chopped down a large tree and built his original cabin around the stump, which the family used as a table. There is no cited source for this information, and it's the only time I've seen that fact reported. I quoted Kettell's article in *Fips, Bots, Doggeries, and More*, and, if that bit of information turns out to be erroneous, I accept the responsibility for perpetuating something that wasn't true. Even so, I can't help but wonder how that fact would get into the record. How long did the original cabin stand? Perhaps it was there long enough for people to tell and re-tell of such a curiosity, if it did indeed exist.

In the 1992 Mount Healthy anniversary booklet entitled *One Square Mile*, Jediah Hill and Henry Rogers are listed as early settlers. The family story is included in the section on local businesses:

> "Mr. Hill came to this wild area in the year of 1819 . . . The late Mrs. Orpha Rogers Blake dated the [covered] bridge at 1850, through talks with her great-grandfather, Jediah Hill's son-in-law Henry Rogers [wrong — Henry Rogers was Orpha's grandfather. Jediah was her great-grandfather] who remembered the building of the bridge when he was a seven-year-old boy [wrong — Wilson Rogers, Orpha's father, was the seven-year-old boy in 1850]."[95]

Updated Version of the Hill and Rogers Family History, 1819–1850

Jediah Stout Hill and his wife, Eliza Hendrickson Hill, both natives of New Jersey, came with their three-year-old daughter, Rachel Maria, to Springfield Township in 1819. Jediah joined his older brothers, Samuel and Charles, and their families, as well as many other friends and former neighbors from New Jersey, and settled on 200 acres of land in Section 28 of Springfield Township that had been purchased by his aged father, Paul Hill. Jediah bought the land from his father for the sum of one thousand nine hundred dollars, according to the deed of sale filed in 1822.

First, Jediah built a log cabin for his family, then commenced to clearing land for farming, and finally sought the advice of his close neighbor, blacksmith and millwright John Lane, to help him set up a sawmill.

The mill, which originally occupied a single-story building with below-ground chambers that housed the water wheel and a pit for collecting sawdust, was likely finished and in operation by the mid-1820s.

Here Jediah turned the logs of the forest into lumber from which he built his barn. He and his family could live in the log house, but they could not shelter their stock nor store their grain without a barn.

With the passing of the years, Mr. Hill's daughter, Rachel Maria, grew into a pre-teen. About this time, Mr. Henry Rogers, the son of a Revolutionary Soldier whose family had migrated to Ohio in 1806, and who had grown up in nearby Mill Creek Township, engaged to work for Mr. Hill.

The 1820 federal census of Springfield Township shows that Henry's married sisters, Elizabeth Rogers McFeely and Jemima Rogers McFeely, both lived on farms near Jediah Hill, so perhaps Henry was already acquainted with his sisters' neighbor.[96]

Four years later, when Henry was twenty-six and Rachel Maria Hill sixteen, they married. It is likely that at least a portion of the Hill homestead was completed in time for their wedding in 1832. Throughout the rest of their lives, the Hill and Rogers families occupied one household.

Henry Rogers and John Lane assisted in the perfection of Obed Hussey's reaping machine in 1835, which may have spurred Jediah Hill to undertake the expenditure necessary to expand his prosperous sawmill to grind flour and meal.

After the 1838 journey back east, during which Jediah and Henry studied how other mills controlled water flow and combined sawing and grinding, they returned to Springfield Township and set about expanding their mill.

Henry and Rachel Maria's only son, Wilson, was born in 1843 and reared in the substantial old homestead on the hill.

The covered bridge over the West Fork of Mill Creek was likely part of the expansion plan, to handle the increased business traffic coming to the mill. Jediah Hill completed the bridge in 1850, and seven-year-old Wilson made the ceremonial first-crossing in his dog cart.

Spoiler alert: Jediah and Henry completed the mill expansion within ten years of their trip east. I'll reveal how I found proof of this in a later chapter.

7
Family Ties to the Abolitionist Movement

Members of the Quaker sect had long professed that slaveholding was incompatible with Christian piety, but they stood almost alone in that belief until after the American Revolution (figures 7.1 and 7.2). The forming of the new country made more Americans consider the slaves' right to freedom as parallel to the colonists' demand for independence.[97]

Even so, the anti-slavery movement in America was largely unorganized until the early 1830s. William Lloyd Garrison founded the newspaper *The Liberator* in 1831, and the following year started the New England Anti-Slavery Society. Though

Figure 7.1. This political cartoon depicts the alliances formed by various political factions who strongly opposed slavery. Image courtesy of the Cornell University Library.

Figure 7.2. Slavery was a divisive issue. Books, pamphlets, and even this calendar were published by supporters on both sides of the issue.

that society grew rapidly, its members split over philosophical differences in 1839. Garrison and his followers lost the support of more moderate members of the society by criticizing churches, denouncing the Constitution as supportive of slavery, and by urging that women hold office within the Society. In 1840, the Liberty Party was founded.

Abolitionists had hoped that either the Whigs or Democrats would take a strong stance against slavery, but this was not to be achieved until the 1850s, when the Free Soil Party and the Whigs, who opposed the extension of slavery into Kansas, formed an alliance out of which the Republican Party was born.[98]

I was aware of Mount Pleasant's reputation as a pro-abolitionist community. As I sought more details about the close associates of the Hill and Rogers families, some interesting connections emerged.

The Free Meeting House that stood on land once owned by Samuel and Mary Hill was the location of several Liberty Party Association meetings, rallies, and conventions in the 1840s.[99]

> **"The Philanthropist**
> edited by G. Bailey, Jr.
> CINCINNATI,
> Wednesday morning, April 21, 1841.
>
> **CONVENTION IN THIS COUNTY.**
>
> We hope our friends in this city and county are looking forward to the Convention in Mt. Pleasant, in this county. It will be held three weeks from to-day. Abolitionists from neighboring counties are invited to attend. . . Let every anti-slavery man in the county, make arrangements so as to be able to leave his business for at least one day — for depend upon it, we shall have a good meeting. The city of course will send up its delegates in crowds.[100]
>
> **LIBERTY CONVENTION ON THE FOURTH OF JULY**
>
> Don't forget the Hamilton County Liberty Convention on the 4th of July next. Mt. Pleasant is but eight miles from the city. We shall have fine speaking there. We hope at least fifty will go from this city. Cars will be running, and the expense be little.[101]
>
> **LIBERTY MEETING AT MT. PLEASANT — HAMILTON CO.**
>
> Meetings in Hamilton County.
>
> The Meeting at Mt. Pleasant on 15th was well attended, notwithstanding the unfavorable weather. The rain poured down in torrents several times during the day. The audiences were addressed during the day and evening, by Messrs. S. Lewis and W. Birney."[102]

Membership in the Liberty Party signaled one's opposition to slavery. Two of Samuel and Mary Hill's daughters married men whose names were listed on the 1841 Liberty Party Roll: Hannah married Dr. Alexander B. Luce and Margaret wed Isaac Lane.[103]

When it came time to gather information on Henry's sisters, I began with Hannah, who had married a man named Zebulon Strong.[104] I chose to research him first because his name was unusual and, I hoped it would be easy to find in census and other public records.

Zebulon Strong was a merchant in College Hill who had also served as a US Postmaster.[105] A Google search associated him with the abolitionist movement. He had built two homes on Hamilton Avenue in College Hill that were reputed to be stops on the Underground Railroad, and one of which is now a charming bed and breakfast inn called Six Acres (Figure 7.3). I read the history of the house on the inn's website,[106] and contacted owner Kristin Kitchen with some additional questions.

She kindly consented to a telephone interview, and I explained the family connections between Zebulon Strong and the Hill/Rogers clan. According to Kitchen, those who were dedicated to the abolitionist movement tended to insulate themselves within their group, even to the point of not marrying outsiders. Individuals who were involved in the Underground Railroad could not risk taking anyone into their confidence who might not share the same convictions and betray them to the authorities.

Accordingly, abolitionists seemed to exist in clusters, and there were many individuals in and around Mount Pleasant and College Hill, the next village to the south, who were passionate about eradicating slavery. Indeed, many of the well-known abolitionists of the time resided in Cincinnati and the surrounding suburbs, including the Quaker Levi Coffin and his wife, Catherine. Coffin was known as the President of the Underground Railroad (Figure 7.4).

Figure 7.3. Six Acres Bed and Breakfast on Hamilton Avenue in North College Hill, once the home of Zebulon and Hannah Rogers Strong, as it appears today.

Others active in the movement in the Cincinnati area included William G. Birney (Figure 7.5), a former slave owner who later became a Liberty Party Presidential candidate and the publisher of the abolitionist newspaper *The Philanthropist*; attorney and future Supreme Court Chief Justice Salmon P. Chase; author Harriet Beecher Stowe; and Rev. Dr. John Witherspoon Scott, an abolitionist professor at Farmers College (Figure 7.6).

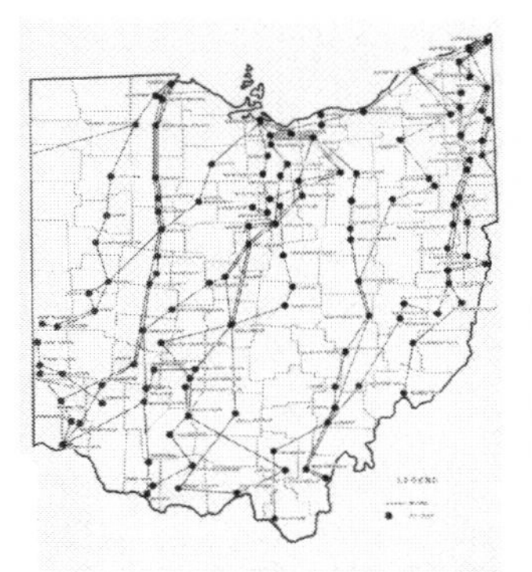

Figure 7.4. This map, which shows routes used by escaping slaves, highlights Ohio's importance in the Underground Railroad movement.

Figure 7.5. James G. Birney, publisher of *The Philanthropist* newspaper and resident of Cincinnati, was the Liberty Party's nominee for president in 1840.

Frank Woodbridge Cheney grew up in Mount Pleasant, where his father, Charles Cheney, was president of the Cincinnati and Hamilton Turnpike Company. Frank and his father frequently transported escaped slaves, which they had hidden under produce in their farm wagon. The elder Cheney hired only tollgate keepers who were sympathetic to the abolitionist cause. The Hamilton Pike was a busy artery of the Underground Railroad, and it was imperative that the gatekeepers be willing to misdirect sheriffs and slave catchers on the trail of runaways.[107]

Figure 7.6. The home of Rev. Dr. John Witherspoon Scott, at 7603 Hamilton Avenue in Mount Healthy, was a place where escaped slaves could seek refuge. A plaque on the south wall of the building attests to its use as a stop on the Underground Railroad.

Mount Pleasant was known as a place where runaway slaves could find help — and also blend in, if necessary. In 1860, there were over fifty free black families residing in Springfield Township,[108] and there were two schools for black children in town.[109]

Abolitionist Connections

John Van Zandt was said to have inspired the character John Van Trompe in Harriet Beecher Stowe's novel, *Uncle Tom's Cabin*. Van Zandt, once a slaveholder in Kentucky, became an abolitionist and conductor on the Underground Railroad in Cincinnati.[110] When he was caught harboring fugitives in the basement of his home in nearby Evendale, Ohio, in the early 1840s, the Sharonville Methodist Episcopal Church (Figure 7.7), of which he was a member, excommunicated him because his anti-slavery convictions were judged to be immoral and un-Christian.

Undaunted, Van Zandt continued to aid escaped slaves as they fled northward to freedom. He was caught again in 1842, and charged with monetary damages by Wharton Jones, a slaveholder who had lost his property. The case, *Jones v. Van Zandt*, challenged the constitutionality of slavery. Salmon P. Chase defended Van Zandt; the Supreme Court ruled against him, saying the federal government had the right and obligation to support slavery.

Jones v. Van Zandt may have hastened the passage of the Fugitive Slave Act of 1850, which carried harsher penalties than the former 1793 Fugitive Slave Act. John Van Zandt went bankrupt after years of challenging his legal case, and died soon after the Supreme Court's decision.

In 2005, the Sharonville United Methodist Church posthumously restored John Van Zandt's membership, with

Figure 7.7. The Sharonville Methodist Church once occupied this building, which still stands at the corner of Creek and Reading roads. In the 1840s, when the Methodist church split over the issue of slavery, Sharonville allied with the southern faction of the denomination. John Van Zandt, friend of Levi Coffin, Harriet Beecher Stowe, and Zebulon Strong, a member of the congregation, was excommunicated because he gave assistance to escaping slaves. Van Zandt's membership was posthumously restored in 2005. Image courtesy of Sharonville United Methodist Church.

a large gathering of Van Zandt's descendants present for the ceremony.

Zebulon Strong was the son of a Quaker mother and a father who had made his fortune selling whiskey to Revolutionary soldiers. Their son served the abolitionist cause with a combination of tenets instilled in him by his parents: the conviction that slavery was wrong and a devil-may-care attitude about following rules.[111]

Strong built two houses that still stand on Hamilton Avenue. Both were known as stops on the Underground Railroad. His children would play in the ravine behind their home, and secretly leave packets of food and other supplies for the fugitives who would hide in the wooded area until nightfall, when it was safe to move to the next hiding place. The Six Acres Bed and Breakfast's website states, "Documents in the Ohio Historical Library speak of [Strong] having a 'false bottom' in his farming wagon where he would pick up his 'passengers' at Mill Creek, which runs along the side of property. He would hide the runaways in the bottom of his wagon and put his crops on the top and take them up to the house for a safe respite before moving them further up Hamilton Pike to the next safe house along the route."[112]

Zebulon Strong was Henry Rogers's brother-in-law. So did that mean Henry was involved in the abolitionist movement, too? Journalist, editor, and historian Jim Blount stated that to best answer that question, we must comprehend the great influence families had over whom their daughters married. Blount expressed his belief that a marriage between Zebulon Strong and Hannah Rogers would not have been possible unless she and her family — including her brother, Henry — were trusted fellow sympathizers.[113]

Isaac Lane, son of the Hill and Rogers families' close neighbor and friend **John Lane**, came from a family of abolitionists. His father's blacksmith shop was reputed to be a stop on the Underground Railroad[114] and his younger brother, Clark, had been forced to leave a blacksmithing job unfinished and flee Hamilton in fear of his life — all because he'd voted for the Liberty Party's presidential candidate in 1844.[115]

Isaac's uncle, Aaron Van Doren Lane, was an Elder of the Christian Church until he was expelled for his abolitionist sympathies. The records hint that church members found Lane's involvement in the Underground Railroad too risky to abide.[116] Church historian D. Mylar Steffy states in his blog that "early records seem to implicate [Elder Lane]" in having tried to use the church building (Figure 7.8) to do what he is suspected of having been doing at his home: providing a refuge for fugitives.[117]

The following excerpt from Steffy's blog gives more detail about Aaron Lane's involvement in the abolitionist movement:

> "In 1848, Mr. Levi Coffin moved his transportation enterprise on the Ohio River from a small town in Indiana to the growing frontier urban center of Cincinnati, Ohio. Several members of the church in Mount Healthy, including the members of the Lane family, provided stations along Mr. Coffin's uncharted

delivery routes. Aaron Lane's stance against slavery arose out of Aaron's own personal experience as a boy, when he was mercilessly abducted and enslaved by Indians and taken to Michigan, only to be discovered several years later by a friend of the Lane family. Aaron's father would bring him home, but the die was cast. No one could argue with his passion, only his sensibility, as his leadership on the Vigilance Committee of the Liberty Party in later years would attest. The example of Mr. Van Zandt, a radical abolitionist who lost everything in 1842 as a result of his zealous abolitionist activities, and the unwelcome price of abolitionism, undoubtedly worked against Aaron as well with the other elders.[118]

Aaron refused to answer the charges against him unless they were written and signed by his detractors. The remaining eldership countered that written charges were not required because all members of the congregation were aware of the charges. Aaron knew that formal written charges could put the church in legal jeopardy, and in true defiance, called the elders' bluff. After several moves and countermoves, including a letter of protest calling for the resignation of the renamed eldership, and his nephew Isaac's own resignation as an elder to join in the protest, the church was left with no other recourse but to disband and leave the matter of Aaron Lane unresolved."[119]

Figure 7.8. The Mount Healthy Christian Church was the center of another abolitionist-related controversy in the late 1840s, which resulted in the expulsion of Elder Aaron Van Doren Lane, friend of the Hill and Rogers families, from the congregation. The church closed its doors for five years over the rift, during which time Lane's nephew Isaac and his wife, Margaret (Hill) Lane worshipped and offered Communion on the front porch every Sunday.

Aaron Van Doren Lane was not the only member of the congregation who was passionate about the abolitionist cause. His name, as well as those of John Lane, Isaac Lane, Clark Lane, and Alexander B. Luce, appears on the Liberty Party Roll for 1841. These men, all of whom have close connections to the Hill and Rogers families, were reputed to be active in the Underground Railroad movement.[120]

Like John Van Zandt, Aaron Van Doren Lane died just weeks after a reconciliation attempt was made. The congregation remained split over the issue, and the church shut its doors for five years, during which time Lane's nephew, Isaac, and his wife, Margaret Hill Lane, worshipped and offered communion on the front steps each Sunday.[121]

Dr. Alexander B. Luce, who had passionately defended Elder Aaron Lane, was a physician and member of the Liberty Party in Hamilton County, and was known to have doctored injured and sick runaways and provided them with warm clothing.

Jediah Hill's niece, Hannah Hill, daughter of Samuel Hill, was married to Alexander Luce in 1834.

Silence and Secrecy

Circumstantial evidence indicates it was likely that Jediah, Henry, and Wilson were sympathetic to, and possibly active in, the abolitionist movement. I'd be proud to know that my ancestors risked their lives and fortunes to aid escaping slaves, and it's hard for someone living in our times to understand why someone who had been involved in the Underground Railroad would be reluctant to admit it, even decades later.

Everything I'd learned about my ancestors led me to believe they had been respectable, moral, Christian people. Though slavery had never been legal in Ohio, being considered a respectable, moral Christian in that time and place did not guarantee that an individual was opposed to slavery. The issue was much more complex than simply living on one side of the Mason-Dixon Line or the other.

Some northern churches, including the Sharonville Methodist Church, which later became the Sharonville United Methodist Church that I attended as a child, and in which I was married, had joined the southern factions of their denominations and preached the morality and necessity of slavery in the decades leading up to the Civil War.

This church that excommunicated John Van Zandt was not alone in its stance that to oppose slavery was un-Christian. In 1836, pro-slavery mobs in Cincinnati attacked the printing press at the office of James Birney's abolitionist newspaper, *The Philanthropist*.[122]

Recently arrived German and Irish immigrants opposed freeing the slaves, because they feared an influx of cheap labor would cut them out of their own jobs.[123]

Though there were many residents of Mount Pleasant with abolitionist leanings, a few miles up the road in Butler County, children got into playground fights over whether they were Yankees or Butternuts, as were known the northerners who stood in sympathy with the South. In Butler County, it was said, "the worst insult a white man could call another white man was 'abolitionist.'"[124]

Tensions ran high in the years leading up to the Civil War, and the divisiveness over the issue did not abate when the war was over. The feelings ran so deep that it was common for people to hide their participation in the Underground Railroad even decades later.[125] Children and grandchildren of those who had been involved in clandestine Underground Railroad activity should have been proud of their forebears, but never heard the stories. Some of the details have been so long buried that they will never resurface. I was able to compare the list of names on the 1841 Liberty Party Roll with the residents of a relative portion of Mill Creek and Springfield townships, and the names highlighted on this 1847 township map had shown their allegiance to the abolitionist cause (Figure 7.9). It should be noted that Dr. Alexander B. Luce owned a home on Compton Road which was known to be a haven for slaves,[126] but because he owned a town lot, his name does not appear on the map. Jediah Hill's and Henry Rogers's names are each marked with black bars as they do not appear on Liberty Party rolls.

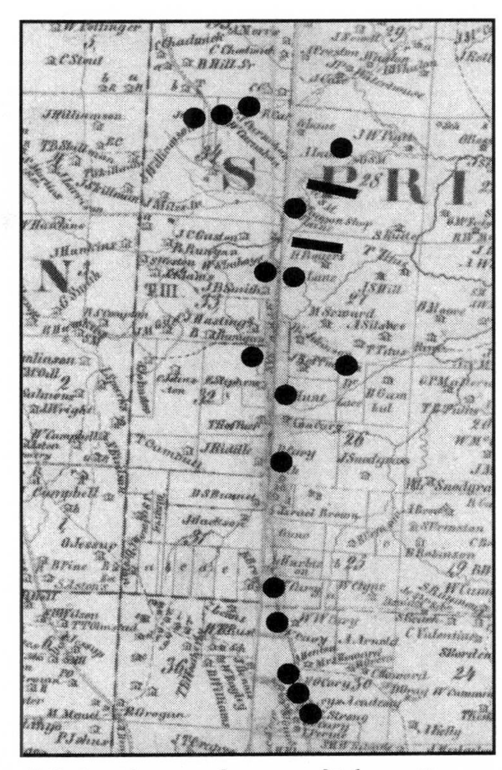

Figure 7.9. Residences of names of Liberty Party members and known participants in Underground Railroad activity have been highlighted with dots on this 1847 map of Mill Creek and Springfield Township. The two black bars identify the properties of Jediah Hill and Henry Rogers.

Sidebar 7.1

In Captivity

A Rootsweb genealogy post had more information about Aaron Van Doren Lane's time as the captive of a Native American tribe. According to a genealogy compiled by Christopher Lane, a descendant of the Lane family, Aaron Van Doren Lane was taken as a child when he was about ten or eleven years old. He was held in captivity for several years, and became accustomed to living with the tribe that had abducted him. When his father came in search of him and bartered for his return, Aaron resisted returning to his family's home.[127]

A traumatic experience such as that would have influenced the entire family's attitude about the damage done when humans hold each other in captivity.

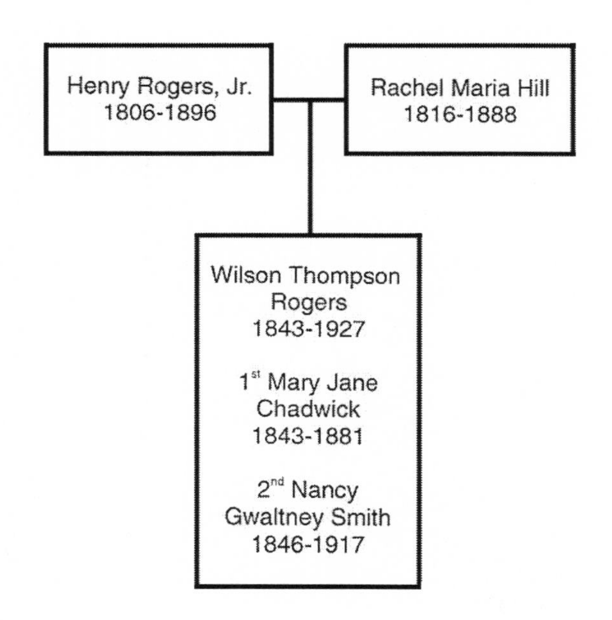
Henry Rogers, Jr.
1806-1896
Rachel Maria Hill
1816-1888
Wilson Thompson Rogers
1843-1927
1st Mary Jane Chadwick
1843-1881
2nd Nancy Gwaltney Smith
1846-1917

8 Wilson Thompson Rogers — Changing Times in the Township

Wilson Thompson Rogers was born on December 29, 1843, sole heir to the legacy his parents and grandparents had built.

It's easy to visualize Wilson Thompson Rogers' early childhood as filled with happy times. Many of the privations that came with pioneer living were no longer a problem in Springfield Township, where Mount Pleasant had grown into a thriving stagecoach stop and business hub located halfway between the larger cities of Cincinnati and Hamilton.

His family attended the Primitive Baptist Church, where he'd been named for the beloved minister,[128] and he trotted down the road to learn reading, writing, and 'rithmetic at the New Burlington School, where his father served on the school board.[129]

Henry and Rachel's only son was one of several boys in the township named for Reverend Wilson Thompson, the popular circuit-riding Baptist minister who visited the red brick church on the northeast corner of Springdale and Pippin roads (Figure 8.1) once a month. Wilson grew up on his family's large, prosperous farm, where he likely swam and fished in the creek, climbed trees, and roamed the fields with neighborhood friends and cousins. The sights, sounds, and smells of his grandfather's mill had always been part of his world.

The bucolic setting was only a short distance from the hustle of the growing town of Mount Pleasant to the south and the smaller hamlet of New Burlington to the northwest. In his later years, Wilson was quoted in *The Story of New Burlington, 1816–1922* that "often as a boy he stood in the

Figure 8.1. The family attended the Primitive Baptist Church at the corner of Springdale and Pippin roads.

stillness of his valley home on winter nights listening to the tinkle of bells on the hames of the horses of the peddlers who were driving in for the night at the Eleven Mile House [tavern at the intersection of Springdale Road and Hamilton Pike]."[130]

The Hills and the Rogers were active supporters of the Baptist church, and as the community grew, so did the opportunities for socializing and fellowship. One year [not specified] the Hills hosted the annual Association meeting, which was attended by followers from Ohio, Indiana, and Kentucky. This account was included in the 1922 booklet:

> "On one occasion, Mr. Jediah Hill was host of the association. The knowledge of the fact a year before made it possible to make ample preparations. The family saved a fine calf and spared no pains to make the plans of entertainment perfect. A procession of carriages a mile long attended the convention and [the occupants] lodged at Mr. Hill's. One hundred and fifty women and children slept in the house during that time while the men rested in the barns. While the grown-ups assembled for worship the children met for pranks and fun."[131]

Wilson's father, Henry, was a strong supporter of the Republican Party, and held the office of Township Trustee for several terms.[132] A township trustee's responsibilities varied with the needs of the community he was elected to serve. These citizen officials could be responsible for budgets for roads and the maintenance of parks and cemeteries, and for assurance that emergency assistance was available for the poor and needy.[133]

In addition to serving as township trustee, Henry Rogers was also a member of the school board. He personally purchased the bell that hung in the second school building constructed to serve the village of New Burlington[134] (Figure 8.2).

Figure 8.2. Henry Rogers purchased the bell shown in the belfry of this New Burlington school building.

At school, young Wilson may have pulled the dark curls and tried to catch the eye of pretty Mary Jane Chadwick, who lived on her family's farm about a mile away down Mill Road.

It's easy to imagine Wilson's mother and grandmother doting on him when he was little, and to envision him being groomed by his father and grandfather as he grew older, so he'd be ready to take over the farm and the thriving mill business.

The first time I saw a photo of Wilson, I said out loud, "Oh, your mama must have been so proud of you." It's easy to see why Mary Jane Chadwick let him catch her eye. Wilson grew into a handsome young man, surrounded by caring family, and heir to everything Jediah and Henry had built.

In the antebellum period, technology was changing fast — faster, in fact, than many small rural mills could afford to upgrade; turbines and circular saws replaced water wheels and the slower sash blade saws. Larger mills were starting to squeeze out the smaller operations, but Jediah's mill's location — close to Mount Pleasant and the surrounding farms — still made it an attractive option for the local trade. Wilson's future looked secure.

The End of an Era

Wilson was ten years old when his grandmother, Eliza Hendrickson Hill, died on June 21, 1854. I have been unable to locate any death certificate, obituary, or will for her. So little is known about women like Eliza, her sister-in-law Mary Wolverton Hill, and her daughter, Rachel Hill Rogers, all of whom should be remembered as pioneers who worked to turn the American frontier into thriving communities.

Jediah's death, on July 4, 1859, was another matter. I was surprised and pleased to discover a transcript of the eulogy delivered by Elder Wilson Thompson included as an appendix to Thompson's autobiography[135] (figures 8.3 and 8.4).

Though the eulogy was over twelve pages long, only a few remarks at the beginning and end referred to the deceased directly:

"A Funeral Discourse
Delivered by Elder Thompson in 1859

Written by Wilson Thompson

THE following discourse was delivered by Elder Wilson Thompson, on the occasion of the death of Jediah Hill, an old and much esteemed brother with whom he had for many years been intimately acquainted, and for whom he entertained the strongest Christian regard and brotherly attachment. It was delivered at the residence of Mr. Henry Rogers, an estimable citizen, near Mount Healthy, Hamilton County, Ohio, on the 31st of July 1859, to a large and attentive concourse of people:

On occasions like the present, when many weeping relatives and sympathizing friends are assembled to drop a tear to the memory of a deceased brother, whose pious life and peaceful death has left so many good examples before them, no subject can be more appropriate than

Figures 8.3. The gravestone of Eliza Hill, located in the Mount Pleasant Cemetery on Compton Road.

Figures 8.4. The gravestone of Jediah Hill, located in the Mount Pleasant Cemetery on Compton Road.

the resurrection of the dead. The importance of this doctrine is second to none in the Christian system of revealed truth.

My habit, on occasions of this kind, is not to say much about the virtues of the dead. On this occasion there is no need of it. He has long lived among you. The hundreds now around me show respect for his memory. His life was the testimonial of his religion; he lived the Christian; his example is before you. He died as the Christian, without a murmur or a fear. He gradually sank down, step by step, for over one year. His pain was not so severe as to make him desire death as a retreat from misery. But with a calm resignation, he submitted all to the will of his God, and without a sigh or a groan, or the distortion of a muscle, he fell asleep like an infant. I have now a vivid recollection when, over thirty years ago, I baptized him and his deceased wife; and from that period to the day of their death, I have always found them sound in the faith and order of the gospel. I believe his neighbors and numerous relatives, many of whom are now before me, will feel a hearty response when I say he lived his religion, and died as he lived, trusting in God, whose service was his delight in life, and whose grace was his solace in death. May we so live, and die, and share the glories of a glorious resurrection. Amen."[136]

It is interesting to note that Jediah and Eliza Hill's baptism, thirty years prior, would have occurred soon after the death of his brother, Samuel.

Inasmuch as Jediah Hill's eulogy summed up his personal standing in the community and his spiritual life, his last will and testament gives clues about his financial life. From that document, we learn that he owned stock in the Cincinnati and Hamilton County Turnpike Company, which, had his wife survived him, she would have been given the use of or dividends from, as well as one hundred dollars per year rent on his real estate. The will also provided that she would have been given one thousand, five hundred dollars in lieu of her dowry, and would also have been allowed to select for herself clothing, personal items, and household furnishings to keep, which were not to exceed fifty dollars in value.[137]

Because Eliza had predeceased Jediah, the Turnpike stock was to be divided between Henry Rogers and Jediah S. Hill, Jediah's nephew, the son of his brother, Charles.

Jediah also bequeathed the sum of fifty dollars each to two young men, Jediah H. Larrison and Jediah H. Hill, on account of them being named after him. Jediah Larrison was the son of Jediah's niece, Rachel Hill Larrison, and Jediah H. Hill, the son of Israel and Harriet Hill.

In his will, Jediah did not specifically mention the mill. The remainder of his estate passed to Henry and Rachel Maria.[138]

A Nation Divided

Throughout Wilson's childhood and teen years, his uncles, cousins, and neighbors were part of the network of people who risked their lives to aid escaping slaves along the routes of the Underground Railroad. There's no way to know if he was taken into the adults' confidence, as was his contemporary Frank Woodbridge Cheney. Cheney used to accompany his father, transporting runaway slaves under produce piled in their wagon, and said in a speech, dated 1903, that his father thought his son's presence made their journeys "to market" look less suspicious.[139]

Just like Frank Cheney, Wilson Rogers came of age during a time of great discord in our nation's history. He was seventeen when the opening shots of the Civil War were fired at Fort Sumter, and, like many of the young men in the community, he joined the local militia and trained to defend Ohio against attack.

Due to its central location in the north and its burgeoning population, Ohio was both politically and logistically important to the war effort. Three hundred twenty thousand Ohio men and boys served in the Union forces, third behind Pennsylvania and New York in manpower contributed to the military.[140]

Henry Rogers, Jr., had not forgotten his father's service in the Revolutionary War. He remained intensely patriotic and a staunch supporter of the new Republican party of Lincoln.[141] Henry allowed passing Union soldiers to camp on his land, and fed them.

The war entered its second, and then its third, year, and Wilson's school days were well behind him. He still had his eye on Mary Jane Chadwick, but he appears to have delayed any definite plans for his future. He was nineteen and a half years old before a major battle was fought north of the Mason-Dixon Line (Figure 8.5).

Within two weeks of the Battle of Gettysburg on July 1-3, 1863, the first significant Confederate threat to the state of Ohio approached, this in the form of Brigadier General John Hunt Morgan. Morgan's Raid was a terrorist attack on the state, meant to draw Northern troops away from their regular duties and strike fear in the hearts of the civilian population.

Harper's Weekly reported in the weeks following Morgan's entrance into Ohio:

> The raid of the rebel Morgan into Indiana, which he seems to be pursuing with great boldness, has thoroughly aroused the people of that State and of Ohio to a sense of their danger. On [July] 13th General Burnside declared martial law in Cincinnati, and in Covington and Newport on the Kentucky side. All business is suspended until further orders, and all citizens are required to organize in accordance with the direction of the State and municipal authorities. There is nothing definite as to Morgan's whereabouts; but it is supposed that he will endeavor to move around the city of Cincinnati and cross the river between there and Maysville. The militia is concentrating, in obedience to the order of Governor [David] Tod.[142]

Figure 8.5. This photograph of Wilson Thompson Rogers was taken in 1863, when he was about twenty years old.

Morgan's Raiders bypassed Cincinnati and General Burnside's troops, and wound their way through the communities north of the panicked city in the late-night hours of July 13. Wild rumor had the force numbered at five thousand, later amended to two thousand, "each one as big as Paul

Bunyan, cruel as Satan, and eager to kill just for amusement." Local contact finally resolved it at the small figure of two hundred twenty-five men[143] (Figure 8.6).

That evening a special militia meeting had been called at Rogers' mill in response to Governor Tod's call to be on the watch for Morgan. After the meeting was adjourned, militia members Wallace Chadwick, Henry Hunt, and Arthur Brackett met a mounted group a little way up the road, and boldly commanded the intruders to halt. Their bravado turned out to be no match for the Raiders' superior numbers, and the three young men were taken prisoner. They were forced to scout the party several miles to the east, where Morgan instructed them that they might go home, as they had no further need of them. But he did need their horses. The young patriots returned from their adventure on foot.[144]

While in the area, Morgan sent spies dressed as country bumpkins into Cumminsville, a village near Cincinnati, where they spread the rumor that Morgan's men were about to attack Hamilton, Ohio. Half the Cincinnati garrison was hurried to Butler County, while the remainder protected the city. Morgan slipped between the two forces and rode east.[145]

Some of the Raiders rode up Colerain Avenue and swung east into New Burlington, two miles north of Mount Pleasant. They stopped for a while in the vale just west of the little church at the corner of Mill and Springdale roads, about three-quarters of a mile from the Rogers farm.[146]

It's likely that Messrs. Chadwick, Hunt, and Brackett returned home as fast as they could to spread the alarm, because local lore tells that when some of Morgan's scouts left New Burlington and penetrated the farmland that circles Mount Healthy, they discovered many "squirrel hunters" armed and eager for contact. The name came from the dress of the armed citizens and the weapons they bore. The squirrel

Figure 8.6. Morgan's Raiders caused widespread panic when they invaded southern Ohio in July, 1863. Some of the Raiders reportedly camped at the Baptist churchyard, less than a mile from the Hill/Rogers homestead. This illustration, from the August 15, 1863 issue of *Harper's Weekly* magazine, depicts Morgan's Raiders entry into the town of Old Washington, in Guernsey County.

hunters were enough of a deterrent, and that group of Raiders pressed on to the east.[147]

By the next day, Morgan's Raiders were miles to the east and the excitement was over. But surely this raid, which struck like lightning then moved on before the thunderclap, left a sobering and lasting impression on Wilson and his comrades-in-arms.

In the aftermath, citizens whose property had been stolen or damaged in Morgan's Raid could file claims for restitution. Cyrus Chadwick, Wallace and Mary Jane's father, filed a claim for the loss of a horse, saddle, and bridle, valued at one hundred ten dollars.[148]

SIDEBAR 8.1

They Renamed it Mount Healthy

The residents of Mount Pleasant were spared from cholera epidemics in 1832 and again in 1849–1850, possibly because Mount Pleasant was situated on rolling hills that provided natural drainage and made water contamination unlikely. There was always "splendid and abundant well water."[149]

The dreaded cholera affected much of the nation, partly because of improvements in infrastructure. Though canals, railroads, and steamboats allowed people to travel more easily than ever before, many of those travelers brought disease with them. Stagnant water in canals allowed cholera to fester.

Eight thousand people in Cincinnati died in the 1849 epidemic, and many fled the confines of the city, ending up in Mount Pleasant.[150]

Sometime in the mid 1800s, the post office required the village to change its name, to avoid confusion with another village of the same name in Jefferson County, near Steubenville. No one seems to know who suggested the name Mount Healthy, but it seems to have been chosen because the village's residents had been spared from the cholera epidemics.

The 1869 Hamilton County map (Figure 8.7) designated the village as both "Mount Pleasant" and "Mt. Healthy P.O."[151]

The name of the town was officially changed to Mount Healthy in 1884, "but it was many years later before the community began thinking of themselves as residents not of Mount Pleasant, but of Mount Healthy."[152]

SIDEBAR 8.2

School Days

Though the national average for school attendance for children aged five to nineteen years of age in 1870 was around fifty-five percent,[153] only about five percent of teenagers attended a true secondary school. High schools were more plentiful in the north than in the south at the time of the Civil War.[154] According to 1860 census data, both Wilson Rogers and Mary Jane Chadwick, the girl who would one day become his wife, attended school as sixteen-year-olds, as did forty-two percent of the boys and forty-five percent of the girls their age in Springfield Township.[155]

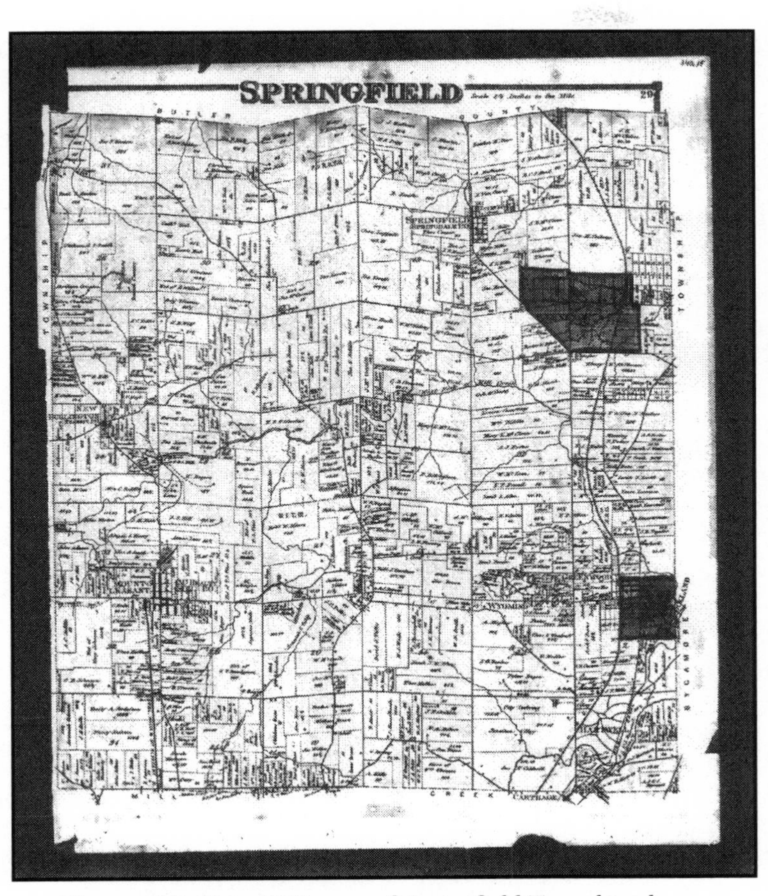

Figure 8.7. This 1869 map of Springfield Township shows the town of Mount Pleasant and also the Mount Healthy Post Office designation.

Sidebar 8.3

Elder Wilson Thompson and the Day the Mississippi River Ran Backwards

Elder Wilson Thompson (Figure 8.8) is regarded as the "ablest Primitive Baptist minister that ever lived in the United States."[156] He was ordained in 1812 and, during his early work in Missouri, he baptized between four and five hundred people. As the area had recently been affected by the New Madrid earthquakes, regarded as the largest seismic activity ever known to the United States, the young minister may have had little difficulty leading the frightened people to salvation.

Three of the New Madrid earthquakes are on the list of America's top earthquakes: the first one on December 16, 1811, a magnitude of 8.1 on the Richter scale; the second on January 23, 1812, at 7.8; and the third on February 7, 1812, at as much as 8.8 magnitude. The upheaval after this third earthquake turned the river against itself to run backwards, created a lake, devastated thousands of acres of forest, and left fissures in the earth five miles long.[157]

Elder Thompson brought comfort to people frightened by the awesome and terrible power of nature that manifested in the New Madrid earthquakes. I wondered how he dealt with the issue of slavery that divided the nation just as the earthquake left fissures carved in the earth.

Though Thompson was a native of Kentucky and lived in Missouri, there are no records that show he ever owned slaves. I looked for evidence of how he led his congregations,

including the one attended by the Hills and Rogers, on the issue of slavery, and found record of an incident that casts light on his feelings about it.

In the story, a slave came to church and asked to be baptized against the will of his master. Elder Thompson questioned the slave about how he intended to serve two masters when their wishes for him conflicted. The slave's answer was that he would obey his earthly master as long as it did not conflict with the will of his heavenly Master. Thompson, who had been threatened with legal action if he dared to baptize the slave, replied, "If you do not fear your master's lash, I do not fear his law." He baptized the man in the creek that ran beside the church.[158]

Figure 8.8. Reverend Wilson Thompson was known as one of the most gifted minsters of his generation, and was revered in the communities he served.

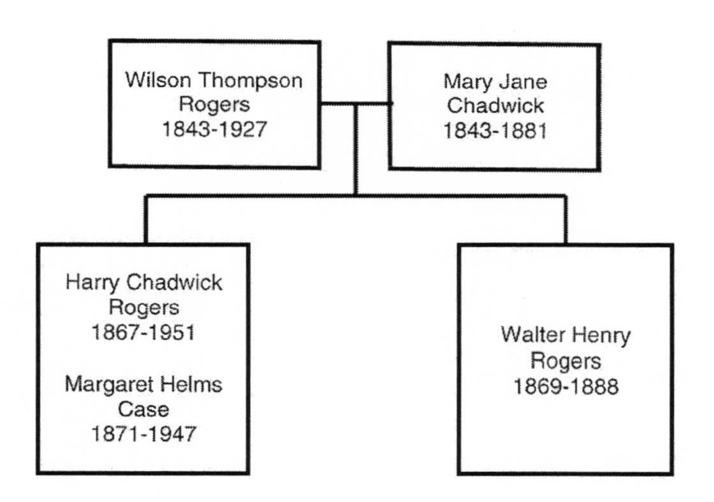
Wilson Thompson Rogers 1843-1927
Mary Jane Chadwick 1843-1881
Harry Chadwick Rogers 1867-1951
Margaret Helms Case 1871-1947
Walter Henry Rogers 1869-1888

9 The 138th O. V. I. Goes Off to War

In 1864, scarce of troops and not ready to launch a second military draft, President Lincoln accepted the state militias from Ohio, Indiana, Illinois, Iowa, and Wisconsin for one hundred days' service in the active military, so to spare the more seasoned troops for the heavy fighting. Ninety-two young men from Springfield Township who were members of the local militia, among them Wilson Rogers and Mary Jane's brothers, Wallace and Darwin Chadwick, began their hundred days' service in the 138th Ohio Volunteer Infantry, Company F, on May 2, 1864.[159] They arrived in Washington, D.C. on May 22 and took up positions in the defenses south of the Potomac River.[160]

Wallace wrote twenty-three letters home to his wife, Rebecca Williamson Chadwick, which serve as a first-hand account of what these young men experienced. Wallace was away from home for the first time, and many of his observations reminded me of Henry's journal entries from his trip east in 1838. Wallace commented on the terrain, the quality of the soil, and the people he met, but it seems he was always aware of the real reason he was there.

"May 18, 1864:

I wouldn't trade forty acres of land in Hamilton County for five thousand acres of this unless I had the means to live without labor. . . I hope I may be permitted to return to old Hamilton County and enjoy its many pleasures and advantages with you and the children, but if, through my small influence there should be something accomplished toward crushing out this infernal rebellion, I shall feel that the government is welcome to what small sacrifices I have made, sacrifices that they were duly entitled to.

July 10, 1864,

I am on duty now. It is extremely unpleasant to be sick out here, yet our regiment is very lucky thus far, not having lost a single man by death.

Through the day we can see the white smoke of exploded shells in the air. On the north side of the river are two regiments of One-Hundred-Day men, one of which has lost two men killed and five wounded on picket duty. It was partly their own fault as they disobeyed orders and fired first, when there was an agreement that there should be no picket firing. Now they are more peaceably inclined and are trading coffee for tobacco with the Johnnies, who say they are on short rations. Some of them come within our lines nearly every day and give themselves up. All say they are tired of it and anxious for peace — all except their officers. With them it is rule or ruin.

A few nights ago I dreamed I was at home. I thought I had Laura [*their baby daughter*] in my hands and was kissing her. Oh, how sweet she laughed and how glad I was! I thought Ellsworth was asleep. Then I awoke. All was a dream yet a happy one, and I hope in another month I may be home.

Give my love to all the family."

Apparently the men of the 138th found time for a sightseeing trip of sorts, and just like soldiers from other wars, Wallace sent home a souvenir to Rebecca.

"Fort Tillinghast, May 23, 1864,

Dear Wife,

We have changed camps once more. We are now situated at the above named place (Figure 9.1). How long we may be here I know not.

Washington did not meet my expectations by any means. There are a few nice buildings, but the majority cannot come anywhere near Cincinnati. I was through the Capitol yard, a nice grove of several acres, all laid off with nice walks. It is a very beautiful place. The Capitol is a splendid affair but I think rather too low for the amount of ground it covers (Figure 9.2). There is one main building with two wings with large pillars cut from solid rock. Many pieces of sculpture are located in different parts of

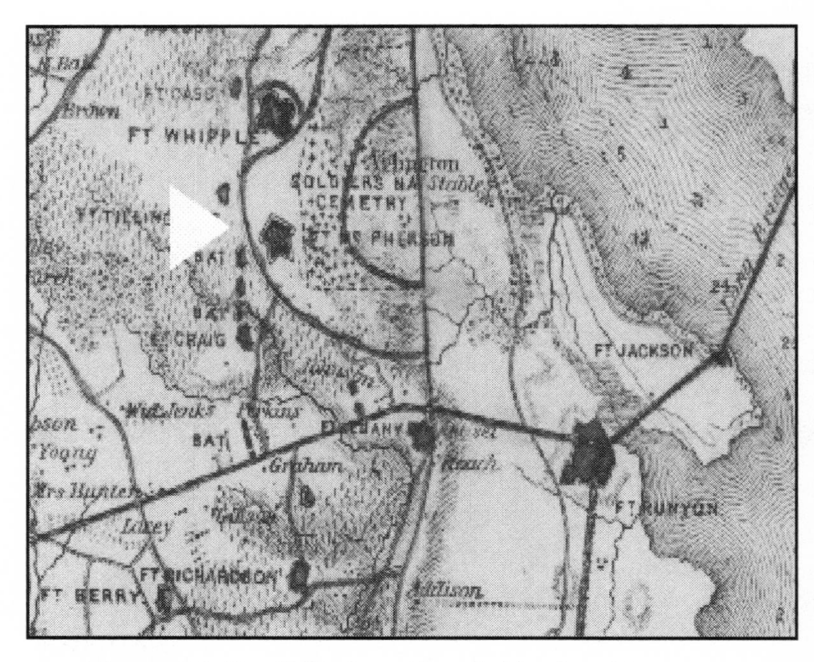

Figure 9.1. Fort Tillinghast, identified by the white arrow, was one of many Union forts located near Arlington, Virginia.

Figure 9.2. The Capitol building was under construction when the 138th OVI was encamped nearby.

the building, some of them very nice, but it will be some time before it will be completed, perhaps years.

We marched through the city and crossed on the Long Bridge to this side, camped in front of the residence and on the lawn of the rebel General Lee (Figure 9.3). It is the most handsome situation I ever saw, commanding a full view of the Potomac for miles up and down the river, in a natural grove, on Arlington Heights. . . The house is of rather an ancient style, having been built in 1817. . .The buildings were planned and built by George Washington Parke Custis, who died in 1837, a descendant of the wife of the Father of Our Country. General Lee came in possession of it in 1837 and because of his traitorism, has thus soon been dispossessed of the finest natural situation I ever saw.

I send you a couple of flowers we pulled from the flower garden as trophies of the home of the rebel general."

The soldiers of the 138th Ohio returned home in July 1864. They had not lost a single man in combat, and they counted themselves lucky as they returned to life on their farms. The war continued, far from home, for another nine months.

Wilson Rogers and Mary Jane Chadwick were married eleven months after the war was over, on March 15, 1866.[161]

Earlier in this book I lamented how little information I was able to find about Eliza Hendrickson Hill and Rachel Hill Rogers. These women were the wives and partners of two of the main characters in my story, and I wish I knew more about them. The details that survive pertain mostly to their marriages, not about their personalities, talents, dreams, or even their daily activities. Henry's travel journal mentions his wife and mother-in-law getting dresses made and having their bonnets trimmed while on the trip east, but that was hardly substantive information about them.

Throughout much of history, a young woman's first duty was to marry; her family often wielded more influence over the choice of husband than she did. Eliza Hendrickson was

Figure 9.3. Stereoscope image of Confederate General Robert E. Lee's home in Virginia.

eighteen when she married Jediah Hill in 1815. She and her sisters, Phoebe, Mercy, and Mariah, all married into the Hill family. Phoebe married Jediah's cousin, Benjamin, and Mercy and Mariah married Jediah's brothers, Charles and David, respectively.

Rachel Maria Hill, Jediah and Eliza's only child, was a mere sixteen years old when she married Henry Rogers, who was ten years her senior. I've often wondered if that marriage came about because Jediah knew he'd found the right man to help him run his farm and his mill, and offered Rachel as part of the deal so Henry would someday inherit the business.

When I expressed to my husband my worry that perhaps Rachel hadn't had a say over her own destiny, he responded, "I'm sure she liked Henry just fine."

The bare-bones information I had about Eliza and Rachel was still more than I knew about Wilson's first wife, Mary Jane Chadwick.[162] My line was descended through Nancy Gwaltney, Wilson's second wife, and when I began my search for Mary Jane, or Mollie, as she was known, she was nothing more than a name on a headstone in the little New Burlington cemetery.

The name Chadwick appeared several times on both the 1847 and the 1869 township maps. Upon close examination, I noticed two pieces of property next to Chadwick land owned by Mrs. M. J. Rogers on the 1869 map. Her parents, Cyrus and Elizabeth Chadwick, were present in the 1850 and 1860 census records. I found records of both of their deaths in 1868 on Ancestry.com, but their causes of death were not listed. Mary Jane's oldest sister, Cynthia, had died in 1860. This piqued my curiosity. What had happened to all of them?

According to Wilson's obituary, he had lived on the family farm his entire life, except for four years, which he spent in Colorado, on "account of his wife's illness."[163] If an illness had prompted Wilson to take Mary Jane to a mountain climate for treatment, it was probably tuberculosis. Was it possible that her parents and sister had also died of the disease? I began to search for someone who might know the answers.

Mary Jane's eldest brother, Wallace, was a main character in the Morgan's Raiders tale. I dug around online to see how much I could learn about him. He'd married Rebecca Williamson, one of the daughters of Amos Williamson, another prosperous farmer whose forebears had migrated west to Springfield Township from Hunterdon County, New Jersey.[164]

A Google search of Wallace's name turned up a question posted on a genealogy chat room by a Sandra Chadwick Mussey in 2002. Fourteen years had passed; was Sandra still out there somewhere, and was she still interested in genealogy?

I located her on Facebook with no trouble, grateful that she'd included her maiden name in her profile. When we made contact a day later, Sandra seemed delighted to hear from me, and we eagerly shared information about our families. She was descended from one of Wallace's sons, who had gone west to California. She'd never been to Ohio, and she appreciated my firsthand descriptions of the ancestral home (Figure 9.4).

Figure 9.4. The Chadwick home on Mill Road in Springfield Township, as it appears today.

She had birth and death dates for all the siblings. Mary Jane's parents had both died of cholera, within months of each other. Mary Jane's sister, Cynthia, had fallen victim to typhoid, and died on the very day that she was supposed to be married to Simeon Williamson.

It was Sandra who told me about the existence of Wallace's letters home to Rebecca, and of course I was anxious to read them. She didn't have good quality images to send me, but suggested I might locate a copy of the journal in which they were published online. My heart sank when I saw that the journal had been published in June 1943, sure that my chances of locating a copy were slim. I crossed my fingers and did an online search. One copy was for sale at a used bookstore in Delaware, Ohio. I couldn't believe it, because I make frequent trips to that lovely town to visit a friend when I'm in Ohio. It seemed like the planets were aligning. I called the bookstore and bought the book.

Sandra had shared a great deal of information, and I thought I'd learned enough to carry on with. Then she sent another email, asking if I had a copy of Wilson and Mary Jane's wedding photo.

What? No, I didn't! Soon after, a scan of a thirty-year-old photocopy of the original 1866 photo appeared in my inbox.

Sandra apologized for the image's poor quality, but at the moment, I didn't care. I was thrilled to see Mary Jane Chadwick for the first time (Figure 9.5). She was dark-haired, winsomely thin, and fashionably dressed in a gown with striped insets. She stood with her hands lightly resting on Wilson's shoulder. He was dressed in a dark suit, holding a walking stick. His hairline had receded a bit in the three years since his other photo was taken, and he now had chin whiskers, no mustache. Mary Jane's girlish silhouette hinted that she might already have been ill at the time of their marriage.

After the wedding, they went to housekeeping and started their family. Harry Chadwick Rogers was born on July 26, 1867, and Walter Henry Rogers joined his brother on September 27, 1869. [165] Anna Strong, Wilson's first cousin once removed and the granddaughter of Wilson's uncle Zebulon Strong, lived with them, presumably to help Mary

Figure 9.5. Wilson and Mary Jane's wedding photo. They were married on March 15, 1866.

Jane with the housework and the children. According to 1870 federal census records, Mary Jane's net worth was ten thousand dollars, compared to Wilson's eight hundred.

The information in that census indicates the little family lived next door to Henry and Rachel Rogers.[166] That dwelling no longer stands on the property, but in a stereoscope photograph taken in the late 1860s, a smaller whitewashed building is visible through the trees between the house and the mill.

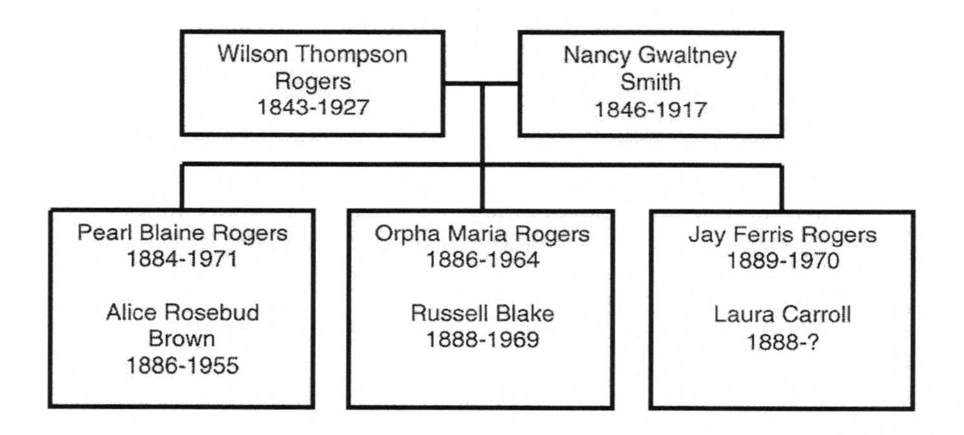
Wilson Thompson Rogers 1843-1927
Nancy Gwaltney Smith 1846-1917
Pearl Blaine Rogers 1884-1971
Alice Rosebud Brown 1886-1955
Orpha Maria Rogers 1886-1964
Russell Blake 1888-1969
Jay Ferris Rogers 1889-1970
Laura Carroll 1888-?

10 Wilson's Post-War Troubles

Henry Rogers assumed sole ownership of the mill after Jediah Hill's death in 1859. Area climatological data indicates that there were droughts in 1862, 1863, 1867, 1870, and 1874, just a third of the years of Henry's ownership of the mill.[167] He had received sufficient rainfall most years, and had time to recover between drought years. There was still abundant timber on their land to be sawed into lumber for sale. We can assume the mill was financially sound when Henry and Rachel Rogers deeded the mill, their 194 acres of farmland, and all their other real property to Wilson on April 15, 1875.[168]

We can assume that when Henry passed the torch to Wilson, he believed that the mill would continue to be profitable. However, there were external factors that could have caused the business to stagnate in the 1870s. They couldn't have known it at the time, but the influx of new settlers into Ohio had begun to slow.

> "As a farming State, Ohio had thus reached maturity, but the Industrial Revolution was only beginning. The glow of furnace and rolling mill, the smoke of factory, the roar of locomotive, the ever-increasing clatter of hoofs on newly paved streets, and the bright, clear light of the coal oil lamp in the homes reveal better than population statistics the changes of the era of the Civil War. The Buckeye State was a young giant who had merely paused to gird up his loins for new tasks."[169]

Ohio's total population, which had increased by around eighteen percent between 1850 and 1860, saw an increase of just under fourteen percent between 1860 and 1870. This was the smallest increase in the state's population in any decade of its history.[170]

> "Without cheap lands to draw settlers as in earlier decades and with industrial development as yet unable to do more than counteract the loss of population due to the westward migration of Buckeyes, Ohio's rate of growth had slowed down until it was

the lowest for two decades of any of the large states. Ohio was still third in the Union in population, but Illinois had become a close fourth and New York and Pennsylvania were drawing farther ahead."[171]

Business may also have been affected by The Long Depression, which began in 1873 and ran through the spring of 1879. The Long Depression halted the strong economic growth in the United States and Europe, which had been fueled by the Second Industrial Revolution in the decade following the Civil War.

The burden of Wilson's worries and responsibilities, both as a business owner and as a husband, increased in the midst of this recession, and his concern over Mary Jane's worsening health may have shifted focus away from the business at a time when it would have been crucial to consider making improvements to stay competitive in the market.

About a year and a half after he inherited the mill and the farm, Wilson made the decision to seek treatment for Mary Jane's tuberculosis in a more favorable climate. He sold two acres of land to Francis Feiter for two hundred dollars on November 11, 1876, possibly to finance the trip and support his family until he could find work.[172]

The 1880 census finds Wilson, Mary Jane, and their sons in Denver, Colorado, where Wilson was working as a clerk in a grocery store and Mary Jane was keeping house. Though there was a question that asked, "Sick?" on the form, the census taker did not indicate that Mary Jane was ill.[173] From this, we can guess that she was not hospitalized, but living at home when the census was taken. In any case, her period of good health did not last. Her disease advanced, and when it became evident she was not going to recover, they returned home to Springfield Township, where Mary Jane died on January 22, 1881, at age thirty-seven.[174]

Tuberculosis

By the beginning of the nineteenth century, tuberculosis had killed one in seven of all people — more than any other illness in history.[175] Evidence of tuberculosis has been found in Egyptian mummies ten centuries old. Though a cure was discovered in the 1940s, tuberculosis is still one of the largest threats to human health on the planet, with one-third of Earth's population currently affected by the disease.

It is no exaggeration to state that tuberculosis affected every family in America in the nineteenth century. It affected old, young, rich, and poor alike, and, as no one understood how the disease was communicated, no one could predict who would be struck down. The uncertainty was terrifying, and people sought explanations — many of which seem silly today, but were based on observation and cause and effect, rather than medical knowledge. Some believed tuberculosis was hereditary. Others insisted being physically beautiful made one more susceptible. It was also theorized that the disease somehow sought out and claimed intelligent and creative people.[176]

In fact, tuberculosis is an airborne virus. Because people

did not understand how the disease was spread, they often unwittingly exposed others to the disease — in particular, young women who nursed ill members of their families. If those young women came down with the disease themselves, there was an attitude of resignation: "Weak constitutions run in our family." [177] Even after diagnosis, there was no way to know how the disease would progress; a patient could suffer with the disease for years, before either getting well or wasting away. There was also the chance that a patient would contract the "galloping" form of the disease and be dead within a few months.

Many novels set in the nineteenth century have a young maiden character that suffers from lingering consumption. It was a popular theme in contemporary literature because the general population could identify with the pathos of the innocent victim of disease, taken too soon from their earthly life. Many of the prominent writers and poets of the day, among them Keats, Poe, Dostoyevsky, and the Brontë sisters, either suffered from the disease themselves or had family members who were afflicted. [178]

In the novel *Jane Eyre*, young Helen Burns suffers from consumption thus:

> "I am very happy, Jane; and when you hear that I am dead, you must be sure and not grieve; there is nothing to grieve about. We all must die one day, and the illness which is removing me is not painful; it is gentle and gradual; my mind is at rest." [179]

While literature of the period would lead a modern reader to believe that consumptives were like Helen, who calmly accepted her fate, in reality, the disease was not painless or romantic. Patients were racked with hacking, bloody coughs, lung pain, and fatigue, and had the sensation that they were coughing themselves to death. [180] Even so, the consumptive maiden became iconic in fashion as well as literature, and physical characteristics evident in tuberculosis sufferers were considered beautiful and desirable in the Victorian era. [181]

Pale skin, dark circles under the eyes, and prominent cheekbones were hallmarks of the tuberculosis sufferer, as were sparkling eyes, blushing cheeks and rosy lips, which were due to a low-grade fever. Many consumptives had silky, baby-fine hair because their hair fell out as a consequence of the disease. The tiny, wasp waist of a young girl who was literally wasting away was replicated by a tightly boned corset. [182]

Some who recovered from tuberculosis credited cold, fresh, air and a dry climate as part of the cure. Beginning in the 1860s, physicians in the eastern United States recommended that their consumptive patients travel to Colorado to regain their health. "Lungers," or people with tuberculosis, flocked to cities like Denver and Colorado Springs, inundating the doctors and hospitals there, who were without sufficient infrastructure and resources to treat them and support their needs. [183]

Other patients went farther west, seeking a return to good health in a desert climate. Albuquerque and Los Angeles can credit their booming growth at the beginning of the twentieth century to advertisements inviting tuberculosis sufferers to settle there and take advantage of the favorable climate and the most modern treatments available.[184]

Dunham Hospital, the first isolation hospital in the country, was founded in Cincinnati in 1879. The hospital was initially meant to treat patients with smallpox, but over the years the facility burgeoned to a one-hundred-acre campus with facilities to treat both adults and children with tuberculosis.[185]

The disease was brought under control after the development of the BCG vaccine, which was introduced in 1921. Later, in the 1940s, antibiotics were used to cure sufferers. Before these treatments were available, tuberculosis made many people who were in an already weakened state unable to fight off the influenza during that epidemic of 1917–1918.[186]

Figure 10.1. Mary Jane's grave is near other members of her family in the New Burlington Cemetery on Mill Road.

Loss and Starting Over

Wilson buried his Mollie in the little New Burlington Cemetery at the intersection of Mill and Springdale roads, in a plot near the graves of her parents, grandparents, and sister (Figure 10.1). The cemetery was about halfway between Wilson and Mary Jane's childhood homes. How often had he passed by that place going to or from courting?

He continued to live with Henry and Rachel (Figure 10.2), who were now seventy-five and sixty-five years old, as he would have needed help with housekeeping and caring for Harry and Walter, who were then thirteen and eleven years old, respectively.

That spring, the rainfall was below average, and after a wet June, just two inches of rain fell in July and August combined. Wilson left no miller's journal or other records behind to piece together what happened next, but the low rainfall that summer may have been enough to both stunt his crops and impede the mill's ability to operate.

In addition to struggling with unfavorable drought conditions, Wilson's mill was forced to compete with the larger, more efficient steam-powered roller mills that had become prevalent in the area. In March, 1881, the Simpson & Gault Company of Cincinnati, which had formerly been known as Straub Mill Company, manufacturer of buhr millstones, had a two-page spread in *American Miller and Processor* advertising roller mill products. In September, 1881, Stillwell & Bierce of Dayton advertised the Odell Roller Mill.[187]

Hydropower and stone grinding were fast becoming obsolete, perhaps faster than Henry and Wilson had anticipated. Small, rural mills like Wilson's were often forced to close, or else go into debt to update their equipment and remain competitive against the steam-powered mills, which could operate without being dependent on sufficient rainfall.

Wilson had a lot on his mind in 1881. I couldn't help but wonder how many months he mourned Mollie before he began thinking about finding a new wife.

Figure 10.2. The mill as it might have looked around the time of Wilson's return home in 1881.

The following year had average rainfall. Perhaps the farm and mill were doing well enough that Wilson was able to turn some of his attention elsewhere. He and Nancy Gwaltney Smith,[188] from Okeana, in Butler County,[189] a schoolteacher with three children of her own, were married on November 29, 1882.

I don't know when Wilson and Nancy met, but I can guess who introduced them. William B. Hill, who was a double cousin to Wilson's mother, Rachel,[190] owned a ten-acre plot in Section 28, just to the north of the Rogers' tract, as well as one hundred ten acres in Section 35 he'd inherited from his father. When he was growing up, his family's next-door neighbors to the west had been Cyrus Chadwick and his family.

William Hill was undoubtedly aware of Wilson's troubles, and he and his wife, Sarah Sater Hill, may have been the ones to introduce Wilson to their niece, Nancy Gwaltney Smith. I'd assumed Nancy was a widow, and was surprised to learn she was a divorceé.[191]

This discovery challenged my modern-day perception of the mores and conventions of the Victorian era on several levels. Did people divorce in the 1870s? I was under the impression that divorce was looked upon as a disgrace in good families, and that it was impossible for a couple to divorce without some scandal or high drama.

Even if Nancy Gwaltney Smith had had ample grounds for ending her marriage, was a divorced woman. . . respectable? Would she have been a suitable choice for a wife? Had

Wilson's judgment been clouded by his grief? I turned my attention to learning more about my great-great grandmother.

Nancy Gwaltney, whose miniature portrait in the Sater family genealogy shows her as an attractive, dark-haired young woman, had married Joseph B. Smith on April 7, 1864, when she was eighteen years old, he twenty-four.[192] 1870 was the first census in which they should have appeared as a married couple, but instead I found Joseph B. Smith living on his parents' farm in Reily Township in Butler County, and working as a field hand. His two daughters, Dixie May and Evaline, then aged five and two years old, were living there with him.

Nancy was nowhere to be found in the 1870 census, and neither was her infant son, Ora, who had been born on April 17,[193] and would have been about two months old when the census taker came around to Morgan and Reily townships in June.[194]

Ten years later, the 1880 census listed Nancy and Ora in her parents' household. Nancy's entry was marked with a D for Divorced.

In contrast, Joseph B. Smith had remarried, and was living nearby in Morgan Township with his second wife, Jane, and their four children, U. Emma, Elva, J. Walter, and Ollie. His eldest daughter, D[ixie] May, now fourteen, was listed as part of his household. Margaret Eveline, who would have been twelve, is absent from both parents' and her paternal grandparents' households in this census. It's interesting to note that Joseph did not mention his divorce to the census taker.

I wondered why Joseph had received custody of the couple's two daughters. And since he had, where was Eveline between the ages of two to twenty-three, when she married?[195] I was never to learn the answer to the second question.

According to the Sater family genealogy, the three children lived with Nancy on her parents' farm, which they may have done, even though that doesn't agree with the census data.[196]

I deduced from the ages of the children in Joseph's second marriage that Joseph and Nancy divorced between 1870–1873, and Joseph had remarried soon after. I called the Butler County courthouse's records room and sought the assistance of volunteer genealogist Bob McMaken. When I joked that there couldn't be that many divorce records to search from the 1870s, he responded that there were "more that you would think."

A few days later, Bob reported back with his findings. Nancy Smith had filed for divorce in January, 1873, listed as Case # 8293 on volume 17, page 135 of the Butler County Common Pleas Journal. The only other information on file indicated that the action had been dismissed, at plaintiff's costs, by consent of the parties.

The petition for divorce was dismissed. So, if Nancy and Joseph were never officially divorced in Butler County, were they bigamists? Bob hastened to say that they could have filed for divorce in any county in Ohio, and perhaps Joseph had filed in one county while Nancy had filed in another. For them both to remarry, the divorce must have been finalized and recorded in some other county. A records search in adjacent counties yielded no information, and so any details regarding Nancy's grounds for seeking the divorce must, for now, remain a mystery.

Divorce, Victorian Style

In the Victorian era, maiden, wife, and mother were all held up to an exalted standard of purity. It was believed that women were governed by emotion and sexuality, rather than logic and reason, and were thereby incapable of supporting themselves. In the years following the Civil War, author and activist Harriet Beecher Stowe campaigned for the expansion of married women's rights, arguing in 1869 that . . .

> "[T]he position of a married woman . . . is, in many respects, precisely similar to that of the negro slave. She can make no contract and hold no property; whatever she inherits or earns becomes at that moment the property of her husband. . . Though he acquired a fortune through her, or though she earn a fortune through her talents, he is the sole master of it, and she cannot draw a penny. [I]n the English common law a married woman is nothing at all. She passes out of legal existence." [197]

Historian Amanda Vickery echoed the same sentiments about the Victorian woman's virtual enslavement to marriage.

> "After a woman married, her rights, her property, and even her identity almost ceased to exist. . . . Indeed it is understandable to see why many women saw marriage as falling little short of slavery. Victorian society viewed marriage as women's natural and best position in life, and men agreed, seeing marriage as an expected duty of women." [198]

I set out to learn about the laws and social conventions of the time as they pertained to divorce, and was surprised to discover that divorce as a civil court action had been around in the American colonies since the early 1600s.[199] I was also surprised to learn that the Puritans viewed marriage as a legal transaction rather than a sacrament of the church, and while that may have made obtaining a divorce easier in some ways, laws that curtailed women's rights made it exceedingly difficult for a woman to survive without a husband.

In the nineteenth century, as they do today, courts outlined provisions for minor children on a case-by-case basis, at the time the divorce was granted. For example, when Patrick and Susannah Martin of Gibson County, Indiana, divorced in 1832, their daughters, Jane and Martha, went to their father while the mother received "the babe, two beds, two colts, and one cow." [200]

The Custody of Infants Act, passed in England in 1839, was meant to award custody of children under the age of seven to their mothers, and older children to their fathers, with custody transferring to the father once a child passed the age of seven. Parliament passed an amended version of the Custody of Infants Act in 1873. This act, known as the Tender

Years Doctrine, stated that young children were assumed to benefit from being in their mother's care until the age of sixteen. Many states in America adopted laws modeled after the Custody of Infants Acts, which gave mothers of unblemished character access to their children in the event of legal separation or divorce.[201]

In Ohio in the nineteenth century, as in our particular case, judges did not always go by the guidelines set in the Tender Years Doctrine, but instead considered both parents on equal footing when it came to deciding who would have custody of the minor children. Children above the age of ten could be allowed to choose which parent they wished to live with.[202]

Around the same time, small, slow gains in the realm of property rights opened the door for women to consider leaving unhappy marriages and envision being able to survive without a husband to support them. In 1846, Ohio passed a law allowing married women to own, but not control, property in their own name, and in 1861, married women in Ohio were granted control over their own earnings.[203]

It is evident that the laws and social conventions of the time made it difficult for a Victorian woman to stand up for herself. The paternalistic society enforced the rules of society that held women to exalted moral standards, as well as the laws and that constrained them. Double standards were commonplace. In the case of infidelity, a woman's affair was enough to end a marriage. Infidelity on the part of the husband, however, had to be coupled with some other offense, like desertion or abuse, to make it a divorce-worthy concern.[204]

Nancy Gwaltney Smith and her husband, Joseph, sought a divorce in a time when amendments to the standing laws had begun to make it easier for a woman to leave an unhappy marriage, but there was still a long way to go. An amendment to the Married Women's Property Act in 1884 made a woman no longer "chattel" but a person, independent and separate from her husband.[205]

Even with the changes in women's property rights and advances in a more equitable system for awarding custody of children, women were still barred from most professions in which they could earn a living wage. If they were no intellectual threat to their husbands, then why was it necessary to hobble them with the male-created and enforced rules?

Though men still held the advantage over women, these small changes led to an increase in divorces. Even so, the statistics are amusingly small when compared to today's divorce rate. In 1890, the first year such data were available, three marriages out of every thousand ended in divorce.[206]

Finally, I wondered, how was an individual determined to be morally fit to be a teacher? I had seen a list of all kinds of things schoolteachers were forbidden to do in 1872 (Figure 10.3), but as it turns out, lists of that kind were fabrications.[207] The Rules and Regulations for the management and government of Common Schools of 1870[208] make no reference to the marital status of the teacher, nor did it specify any rules relating to morality or character.

None of this was what I expected to learn. I was glad Nancy had been able to leave her unhappy marriage and, at least partially, support herself by teaching. In any case, she must have been regarded as suitable for the job. According to the 1880 census data, Nancy's younger sister, Orpha, was also teaching school at that time.[209]

Letting Go of the Legacy

Nancy and Wilson's marriage added at least two more mouths to feed to his household. I assume that, as a married woman, Nancy would have ceased to work outside the home, and would have instead helped Rachel run the household and care for the children. Wilson, now shouldering the responsibility for his aging parents, his wife, and their children, continued to work in the mill and on the farm. But before long, the once-golden boy would face another crisis.

Local lore says Wilson sold the mill in 1887, but gave no details as to why. Early in our collaboration, Steve and I had wondered about, and discussed possible reasons why, Wilson might have sold the mill.

Steve figured he was in financial trouble. But, I had protested, I had found no evidence to support anything of the kind. If I was correct, Steve responded, then he thought Wilson was a spoiled brat. How dare he sell his grandfather's business? It had been Jediah's dream — and his life's work.

1872 Rules for Teachers

- Teachers each day will fill lamps, clean chimneys.
- Each teacher will bring a bucket of water and a scuttle of coal for the day's session.
- Make your pens carefully. You may whittle nibs to the individual taste of the pupils.
- Men teachers may take one evening each week for courting purposes, or two evenings a week if they go to church regularly.
- After ten hours in school, the teachers may spend the remaining time reading the Bible or other good books.
- Women teachers who marry or engage in unseemly conduct will be dismissed.
- Every teacher should lay aside from each pay a goodly sum of his earnings for his benefit during his declining years so that he will not become a burden on society.
- Any teacher who smokes, uses liquor in any form, frequents pool or public halls, or gets shaved in a barber shop will give good reason to suspect his worth, intention, integrity and honesty.
- The teacher who performs his labor faithfully and without fault for five years will be given an increase of twenty-five cents per week in his pay, providing the Board of Education approves.
- You may ride in a buggy with a man, if the man is your father or your brother.

◀ **Figure 10.3.** Lists such as this one have never been proven to be authentic. Rules and regulations for common schools in the 1870s did not specify how to judge a potential teacher's morality or character.

Steve was angry that Wilson had given it up after he'd owned it for what would have been a mere twelve of the sixty years the mill had been in the family.

Steve's assessment of Wilson's callousness and disregard for his family's legacy was always in the back of my mind. Did he really sell the mill without regrets? Didn't he fully appreciate what it had taken for his grandfather to build the business up from nothing?

But as I uncovered the details of Mollie's death, it seemed logical that Wilson, overwhelmed and exhausted by the personal tragedy, didn't want to run the mill any more. Maybe he'd decided that, under the circumstances, a job with a steady paycheck, like the grocery clerk job he'd had in Denver, suited him better than being a business owner. I had already learned that when he grew older, Wilson had worked for the post office.

Late in my research, I discovered a discrepancy. The local histories all indicated that Wilson sold the mill to Charles Hartmann around 1887, but some of the Hartmann family documents, including Charles' unpublished journal, put it at 1883. I requested copies of the documents that recorded the sale of the mill from the Hamilton County Courthouse.

The deed of sale was titled *Wilson Rogers per Assignee "Deed" Chas. Hartmann,* and confirmed Steve's initial theory that the business had fallen into debt. On May 4, 1883, Wilson and Henry each, in separate documents, made assignment of "all his property including the real estate described below, in trust for the benefit of his creditors, to John L. Riddle, Esq., which assignment was filed in Probate Court on the same date. In September, 1883, Charles Hartmann purchased the mill and the three acres of land on which it stood for the sum of one thousand dollars.

I gasped aloud as I read, and over my left shoulder, I could swear I felt someone sigh. A deep sadness swept over me, and I cried, certain that whoever was looking over my shoulder had wanted to shield me from the truth; I found comfort in the fact that the family was able to retain the homestead and the farm.

After Henry and Wilson sold the mill, Wilson "actively engaged in agricultural pursuits in Springfield Township, Hamilton county, Ohio, where he was considered one of the most progressive and well-informed agriculturists."[210]

It was a time for new beginnings. Wilson and Nancy's first child together, my great-grandfather Pearl Blaine Rogers, was born on June 19, 1884.

11 Rogers Wrap-Up

After Wilson sold the mill, the Rogers family continued to reside in the big house. Wilson and Nancy's family expanded to include Orpha Maria, born in 1886, and Jay Ferris, born in 1889. As his young family grew, Wilson's parents and his sons from his first marriage were, one by one, making their exits from his life.

Wilson and Mary Jane's younger son, Walter Henry Rogers, died on January 16, 1888, at the age of eighteen. No death certificate was found, and his cause of death is unknown.[211] He was buried in the New Burlington Cemetery, near his mother (Figure 11.1). I've often wondered if Walter had also contracted tuberculosis, or if he succumbed to one of the other diseases that had claimed members of his mother's family.

Rachel Hill Rogers died a few months later, on April 25, 1888, at age seventy-two. Again, no death certificate was found, and no cause of death is known.[212] Henry Rogers died December 1, 1896, at the age of ninety. The cause of death listed in his obituary was "paralysis."[213] Henry and Rachel were buried in the Mount Pleasant Cemetery on Compton Road (figures 11.2, 11.3, and 11.4), near the graves of Jediah and Eliza Hill.

Wilson's remaining son from his union with Mary Jane, Harry Chadwick Rogers, was married to Margaret Case in 1891, when he was twenty-four and she was twenty. Their

Figure 11.1. The cause of young Walter Henry Rogers's death is unknown.

Figure 11.2. Rachel Rogers is buried in the Mount Pleasant Cemetery, near her parents and husband.

Figure 11.3. Henry Rogers at age 90.

Figure 11.4. Henry Rogers is buried with his wife and her parents near the Compton Road entrance to Mount Pleasant Cemetery.

three children, Grace Ramona, Mary Jane, and Raymond Case, were all born in Mount Healthy (Figure 11.5). The family moved west to Kansas City, Missouri, around the turn of the century.[214]

Wilson and Nancy Gwaltney Rogers' family portrait (Figure 11.6) shows an established family, in contrast to young Wilson and Mollie's wedding photo. Wilson and Nancy are now in their late fifties or early sixties. Wilson's hairline has receded; a gold watch chain loops across his vest. Nancy wears spectacles and has a lace frill at her collar. She looks matronly and aristocratic. Their son Pearl, a young man of maybe twenty, stands at his father's right. Orpha, probably eighteen, is in the center of the frame, a dark-haired beauty with a fashionable Gibson-girl pompadour and deep-set eyes. Young Jay, about

Figure 11.5. Harry Chadwick Rogers and his wife Margaret "Maggie" Case Rogers brought their daughters home for a visit. Back row: Harry Rogers, Alonzo Case, Wilson Rogers. Front row: Maggie, daughter (either Mary Jane or Ramona Grace), Orpha Rogers, and daughter.

fifteen, has scuffed shoes, but his collar and tie are neat and crisp. The family looks like the epitome of turn-of-the-century prosperity and respectability.

Census records had provided me with a great deal of information about the different generations of the family. The information provided by Wilson and Nancy in their final three interviews with the census taker shows some discrepancies and omissions, mostly by Nancy.

In the 1900 census, Wilson reported he owned 120 acres of land and the dwelling on it, but the property was mortgaged.[215] He reported his occupation as postal carrier. I was sorry to see the effects of Wilson's financial troubles from the preceding decade were still a concern, as he had mortgaged the home that had belonged to his family for nearly seventy years.

Nancy Rogers reported that she was the mother of only three living children. I can only speculate why she might have

Figure 11.6. The Wilson Rogers family, late 1890s. Left to right: Pearl, Wilson, Orpha, Nancy, Jay.

Figure 11.7. Wilson Rogers with his mail wagon.

neglected to give the census taker information about her divorce and her three children from her first marriage.[216]

In the 1910 census, Wilson reported owning a mortgaged home, but this time reported it was just a home, and not a farm.[217] This time, Nancy reported that she was the mother of six living children.[218]

In 1920, Wilson's final census, his home was still mortgaged.[219] According to *Memoirs of the Miami Valley*, Wilson was remembered in his later years as "living retired on his beautiful farm, surrounded by his children and grandchildren, loved and respected by all who know him. He was rural mail carrier out of Mount Healthy for over seventeen years, and his work was a model of fidelity to duty" (Figure 11.7).[220]

Obituaries of Nancy and Wilson Rogers

Nancy Rogers died at home on March 25, 1917, at the age of seventy-one. Her death certificate lists a cerebral hemorrhage as the cause of death.[221]

> "Nancy Gwaltney Smith Rogers, daughter of the late James and Sarah Gwaltney, was born Jan. 15th, 1846 and departed from this life Mar. 25, 1917, aged 71 years, two months and ten days.

She is survived by her husband, Wilson T. Rogers, one son and two daughters from her first marriage; Dixie May McCoy, of Hamilton, Ohio; Eva Walling of Okeana, Ohio; and Dr. O.J. Smith ofVenice, Ohio; Two sons and one daughter by her second marriage: Pearl Rogers, ofWinton Place, Cincinnati, Ohio; Jay F. Rogers of Glendale, Ohio and Miss Orpha Rogers. She also left ten grand-children and two great-grandchildren.

In early life she united with the Primitive Baptist church and remained faithful until death.

Thy gentle voice is silent now
Thy warm true heart is still
And on thy noble brow,
Is resting Death's cold chill.

We miss thee from our home, dear Mother,
We miss thee from thy place,
A shadow o'er our life is cast
We miss the sunshine of thy face.

We miss thy kind and willing hands,
Thy fond and earnest care,
Our home is dark without thee
We miss thee everywhere.

Servant of God, well done!
Thy glorious warfare's past,
The battle's fought, the race is won,
And thou art crowned at last."

After Nancy's death, Wilson's daughter Orpha dedicated herself to caring for her father. Wilson Rogers outlived Nancy by ten years, and died of stomach cancer on May 13, 1927, at the age of eighty-three[222] (figures 11.8 and 11.9).

"VETERAN

of Civil War Passes On. Wilson Rogers had Lived in Mt. Healthy Over Eighty Years.

The passing of Mr. *[Wilson]* Thompson Rogers at his home last Friday morning has removed from the Mt. Healthy community another of the old pioneer citizens. Mr. Rogers was well known in the community and familiarly known as "Dad Rogers."

The funeral was held at the residence just north of town on Monday afternoon. Rev. C. O. Cossaboom

Figure 11.8. Nancy and Wilson Rogers are buried in the New Burlington Cemetery.

Figure 11.9. Wilson Rogers' gravestome in the New Burlington Cemetery.

of the Christian Church officiating. Ex-mayor Alexis Brown, who was a comrade of Mr. Rogers in the Civil War, spoke very touchingly of the passing of his comrade leaving himself as the only surviving member of the Mt. Healthy Post of the Soldiers and Sailors Organization. The Masonic service was conducted by members of McMakin Lodge No. 120.

Wilson Thompson Rogers (named after a noted Baptist minister, Wilson Thompson) was the son of Henry Rogers and Rachel Hill Rogers. He was born on December 29, 1843 at the home where he has lived continuously (with the exception of four years) for more than eighty years. He was united in marriage with Mary Chadwick, schoolmate of his at the old Burlington School, which they both attended. To this union was born two children one of whom, Harry Rogers now resides in Kansas City Mo. After four years in Colorado on account of his wife's illness, he returned to the old home, where Mrs. Rogers soon died.

Mr. Rogers was married later to Nancy Gwaltney Smith. To them were born three children, Pearl, Orpha and Jay, all of whom are living.

He leaves to mourn their loss, Harry Rogers, Kansas City, Mo., Pearl Rogers, Reading, Ohio, Jay Rogers Glendale Ohio, one steps-son, Dr. O. Smith, a daughter, Miss Orpha Rogers, Mt. Healthy and a step-daughter, Mrs. Elmer Crouse, Hamilton, Ohio. Fourteen grandchildren, Mrs. Grace Lucas, Springhill, Kans., and Mrs. Mary Teter, of Kansas City, Mo., Raymond Rogers, of Richardson Springs, Calif., Walter, Vernon, [Ruth] Ray, Florence, Winona Rogers of Reading, Ohio. Maynard and Dorothy of Glendale. Dr. Paul Smith of Cincinnati, Mrs. Carl Eberling of Hamilton, Ohio, James and Francis Eberling of Hamilton, Ohio, and six great-grandchildren.

Mr. Rogers was of that race of fearless pioneers who first cleared the forests and built the roads and bridges, and he, with his own hands, did much to make this neighborhood more habitable.

The spirit of enterprise, of optimism, of faith, of his pioneer ancestors, stayed with him through life.

When the days were darkest he believed that the sun would soon shine again.

No condition seemed so hopeless but he claimed that things were already growing better, and though at times his faith might seem to falter, hope never died.

When the Civil War came he enlisted with the 138th O.V.I. and served with his regiment around Richmond, Va. Later in life he enlisted in the U.S. Mail Service and was on duty until a few years ago when, on account of age and faithful service he was honorably retired and pensioned.

Patriotism was a dominant principle in Mr. Rogers' life. He enlisted in the army because he felt his country needed him. As a member of the United States Mail service he performed his duties faithfully and well proud of the fact that he was a servant of the land he loved, and in his long peaceful private life no government ever had a more loyal subject.

He believed in his country and was ever ready to defend it in peace or war.

In our Decoration Day exercises, rain or shine he was seldom absent. He came not because of his pride in being a veteran but with joy that he could do his bit.

When in our future patriotic celebrations we shall miss him, we shall know that though "his body may be mouldering in the grave, yet his soul goes marching on."

It seems most appropriate that in the old churchyard, in which he has always taken so much interest and for the upkeep of which he has labored so faithfully that he be laid to rest, so near his old home amid the scenes that through a long lifetime he has loved so well, "till the morning break and the shadows shall flee away."

Continued Family Connections

Wilson's daughter Orpha was the last family member to own the house built by Jediah Hill. After her father's death, Orpha married a widower, Russell Blake, and continued to reside in the old house on Covered Bridge Road until her death in 1964 (Figure 11.10). She was on friendly terms with the subsequent owners of the mill. My grandmother and her siblings, and then later my father and his brother, grew up in nearby Lockland and Reading, and paid many visits to the farm as children (figures 11.11 and 11.12).

Wilson's death in 1927 ushered out the era of the pioneer; there was no one left in the community who could remember when Jediah Hill ran a small sawmill on Mill Creek near a village called Mount Pleasant. The twentieth century brought with it technological advances that made looking back both impractical and impossible. The subsequent generations of men who owned the mill would make improvements that allowed it to operate as a modern factory and thereby prolong its usefulness to the community.

Figure 11.10. Orpha and Russell Blake.

Figure 11.11. Pearl and Alice Rogers, around 1910. The mill is visible in the left background.

Figure 11.12. The grandchildren of Wilson and Nancy, Easter 1918: Vernon (holding Florence), Ruth, Roy, and Walter.

Sidebar 11.1

An Enduring Gift

When my grandmother, Ruth Rogers, was a little girl, Mrs. Carrie Groff gave her a bisque doll (Figure 11.13). The doll has the striking features and heavy brow characteristic of the French and German dolls popular in the late nineteenth and early twentieth centuries. She has survived the decades with little more damage than a haircut, likely suffered at the hands of my young grandmother. The doll's head is bisque, unglazed china that was favored for its lifelike appearance, and is topped off by the shorn wig made of mohair. Her composition body was molded from a mixture of water, sawdust, and glue, and jointed with ball joints.

In the 1960s, my mother made new clothes, underwear, and shoes to replace the doll's tattered original costume.

A recent appraisal identified my grandmother's dolly as a German Kestner Character Doll, model 143, and dated her at 1909-1910. After the appraisal, the ladies at the Mid-Ohio Historical Doll and Toy Museum in Canal Winchester, Ohio, re-strung her joints, cleaned her face, and pressed her gown, and she's never looked sweeter.

Figure 11.13. Mrs. Carrie Groff gave this doll to my grandmother, Ruth.

Figure 11.14. Pearl and Alice Rogers with their grandchildren. My uncle, Gary Stone, is at center, and my father, Todd, is the baby in the white blanket on Pearl's lap.

Sidebar 11.2

It Only Takes a Spark

My uncle, Gary Stone (Figure 11.14), recalled that his grandfather, Pearl Rogers, took him to visit the mill when he was nine or ten years old, which would have been around 1948–1949. Before he would take him inside for a tour, Pearl checked the bottoms of Gary's shoes, worried that a stray nail or a tack embedded in the sole of Gary's shoe might create a spark, ignite the flour dust, and cause an explosion or a fire in the mill.

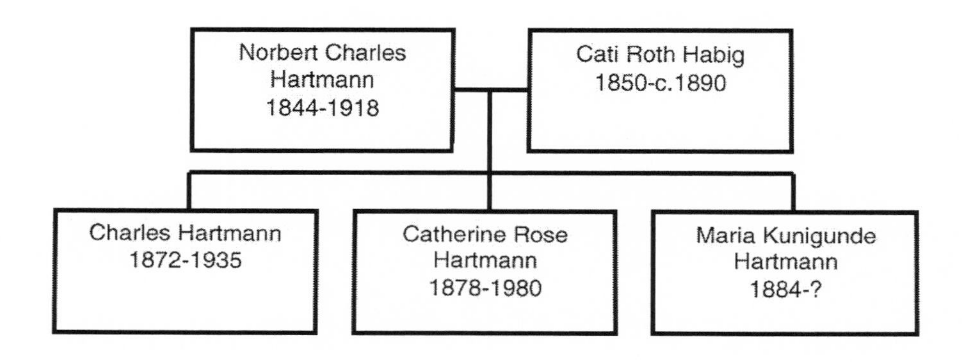
Norbert Charles Hartmann
1844-1918
Cati Roth Habig
1850-c.1890
Charles Hartmann
1872-1935
Catherine Rose Hartmann
1878-1980
Maria Kunigunde Hartmann
1884-?

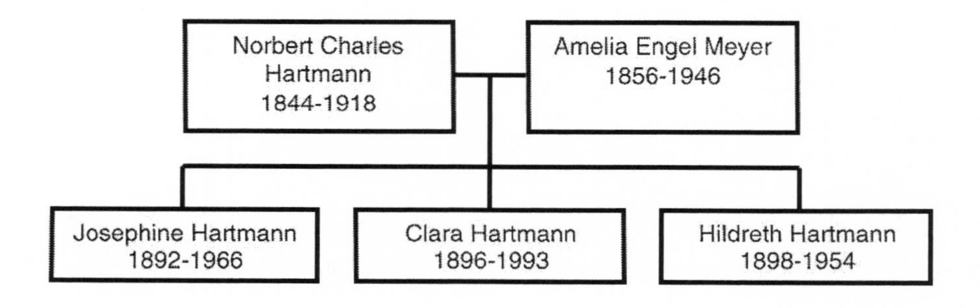
Norbert Charles Hartmann
1844-1918
Amelia Engel Meyer
1856-1946
Josephine Hartmann
1892-1966
Clara Hartmann
1896-1993
Hildreth Hartmann
1898-1954

12
Charles Hartmann — Bread and Family

In the early years of the industrial revolution in Germany, manufactured goods became available at much lower prices, and local cottage industries suffered. Westphalia in the North Rhine was particularly affected, and this caused many people from the region to leave their homeland and seek better opportunities in America.[223]

Between 1860 and 1869, around thirty-five percent of all immigrants to the United States were from Germany.[224] Among them were Norbert Charles Hartmann and his three brothers.

Charles, as he was called after he came to America, was born December 17, 1844, in the village of Dalhausen in North Rhine-Westphalia. His family owned a flourmill, and as it was the custom in the region for the oldest son to take over the operation of the family business, younger sons had the option to either work for their brother, or to apprentice with another family to learn a trade. Charles was apprenticed to a cabinetmaker, and when he was twenty-two years old, he and three of his brothers traveled to Bremen, where they set sail for America on July 20, 1867. They arrived in New York on August 3 and journeyed by rail to Cincinnati, where they had friends. One of the brothers died soon after arriving in America. Charles' other two brothers, John and William, purchased a flourmill in the town of Plaineville, on the east side of Cincinnati.

It is not known where Charles first worked after his arrival in Cincinnati. He married Cati Roth Habig on November 18, 1869, when she was nineteen and he twenty-four. Five of their eight children died in infancy or early childhood. The surviving children were the second-born, John Charles (known as Charlie), the fifth, Katherine Rose (Katie), and the seventh, Maria Kunigunde (Mayme).[225]

Cati suffered from tuberculosis, and the couple was advised to move away from the city in hopes of improving her health. Charles purchased the Mount Healthy Mill from Henry and Wilson Rogers for one thousand dollars in September, 1883.[226] At first, the Hartmanns lived in rooms on the upper floors of the mill, and their two youngest children were born there.[227]Their seventh child, Mayme, was baptized at Assumption Church in Mount Healthy in 1884. Charles' journal ends when their eighth child was born on June 1, 1886.[228] Charles' granddaughter, Millie Hartman, stated that both Cati and the child died soon after.[229]

Charles married a second time, this to Amelia Engel Meyer, on November 21, 1891, and together they had three daughters and one son. The son died in infancy. Josephine, the eldest daughter, was born while the family was living in the mill. Charles built a frame house nearby (Figure 12.1), and the family had moved in by 1896, when Clara was born. Hildreth, their youngest child, was also born in the house in 1898.[230]

The local histories state that about this time, Hartmann phased out the sawmill part of the business and converted the mill from waterpower to steam. The histories cite a shortage of timber as the reason. The rise of larger lumber mills in the area may have also influenced his decision.

Charles may also have experienced some of the same problems with drought and lack of sufficient waterpower that had plagued Jediah, Henry, and Wilson from time to time during each of their ownerships of the mill. Cincinnati was a leading manufacturer of steam engines. In 1898, Charles decided it was time to upgrade the mill, and stop relying on the unpredictable water flow from Mill Creek (Figure 12.2).

During the conversion, the water wheel and turbine would have been removed. The wall separating the water wheel's chamber and the sawdust pit would have been taken down to make way for the steam engine and the drive belts that powered the gears that ran the new roller mills.

Figure 12.1. Charles Hartmann built this house for his family around 1896, before the mill was converted to steam power. Note the picket fence that separates the yard from the mill race's ditch in the background. Pictured, left to right: (back row) Charles, Jr., Katherine, Amelia (front row) Josephine, Clara, Mayme.

Lumber had always been a reliable source of income for the mill, but, as the local timber supply dwindled, Hartman decided to phase it out, and focus on flour milling. Census data provides clues that show the transition occurred between 1900 and 1910. In the 1900 census, Hartmann told the census

Figure 12.2. The mill, after its conversion to steam power in 1898.

taker that he was the owner of a saw and gristmill. His son, Charlie, then twenty-seven years old, was employed as a sawyer in the mill. Fred Bishop, a neighbor, reported his occupation was "engineer in saw mill."[231]

According to the 1903 Annual Report of the Department of Workshops, Factories, and Public Buildings, Charles Hartmann employed five men in his mill, which was the only mill in operation in Mount Healthy.[232]

In the 1910 census, Charles Hartmann's occupation is listed as flour mill.[233] The Hartmanns' entry included a hired man as a member of their household, who worked as a teamster for the mill.[234]

Roller mills produced the highest quality flour available to date, and Charles was eager to use the upgraded equipment to its best advantage. Rather than grinding for local farmers in exchange for a portion of the grain, as was common practice for small, rural mills, he chose to focus on milling and marketing his own product, and for that, he needed a recognizable brand name. Soon, he was selling Pride of the Valley Flour to businesses in the Cincinnati area, including the Gibson and Sinton hotels (figures 12.3 and 12.4).

Picnic Grounds and Spas

Beer gardens developed in Germany in the nineteenth century, partly because the risk of fire from the brewing process had led authorities to limit beer brewing to the cooler times of the year. In order to provide beer to thirsty customers during the hot summer months, large breweries dug cellars in the banks of rivers to store beer, and then covered the riverbanks with gravel and planted trees to provide cooling shade. Soon, beer cellars evolved into beer gardens, with simple tables and benches set up among the trees. They became popular venues for outdoor socializing.[235]

It's easy to see why factory workers who spent their days indoors, often in poorly ventilated buildings, flocked to open-air beer gardens and other types of amusement parks.

Figure 12.3. Hartmann marketed Pride of the Valley flour to bakeries, restaurants, and hotels in the Cincinnati area, including (A) the Sinton and (B) the Gibson.

Charles Hartmann's brothers, John and Will, purchased property that included a working mill from the heirs of Nathaniel Armstrong, a pioneer who settled near the present-day village of Indian Hill. John and Will Hartmann ran the mill and operated a picnic ground known as Hartmann's Grove, which had open grass areas for picnics, a dance hall, and boat rentals. City dwellers took the Traction Rail Streetcar and Pennsylvania Railroad to the picnic grounds to escape the stifling summer heat (figures 12.4 and 12.5). The Hartmann brothers operated the grove for

Figure 12.4. Hartmann's Mill at what is now Avoca Park, as it appeared in 1902.

Figure 12.5. The boating lagoon, made from the tailrace at Hartmann's Grove.

twenty-nine years, and were pioneers of the picnic basket park in Cincinnati. Individuals, clubs, and organizations held outings and picnics there.[236]

The two Hartmann families each had ten children, some of whom played musical instruments, and provided musical entertainment to the picnic crowds on Sundays. John's son, George Frank, ran the boat rentals.[237]

The families sold Hartmann's Grove to the Avoca Park Company, which built amusement attractions, in 1907. The flood of 1913 removed many of the historic buildings on the site, though some of the foundations remain. The disabled veterans built a number of cottages on the property in 1922.

Avoca Park is located at 7949 Wooster Pike/US 50, one and a half miles east of Mariemont and west of Terrace Park. It is the southern trailhead of Little Miami Scenic Trail, which goes north along the Little Miami River. Owned by Great Parks of Hamilton County, the 65-acre area is now being preserved without development except for parking and restrooms.[238]

While Charles' brothers kept busy with their milling and entertainment business enterprises, Charles considered adding a medicinal spa to his own milling business, and had the well water analyzed.

> "Cincinnati, November 17, 1899
>
> Mr. Chas. Hartmann, Mt. Healthy, O.
>
> Dear Sir: The preliminary analysis of your mineral water shows that it contains about 1995 grains of mineral salts to the gallon. These salts consist mostly of Chlorides, some Carbonates, traces of Sulphates, and a comparatively good amount of Bromides. The mineral part is predominantly Sodium (Chloride of Sodium, etc.), Calcium, Magnesium, Potassium.
>
> This water must be classified as a Saline Water of such strength that it has to be ranked rather as a medicinal brine.
>
> Very respectfully,
>
> Dr. W. Dickore."

In *One Square Mile,* there was a photo of people gathered near the well.

"People came to draw water from Mr. Hartmann's well near the mill, which he had dug with the thought of tapping into an ordinary water source [Figure 12.6]. People found the water contained medicinal properties, and word spread until Mr. Hartmann considered bottling the water for sale. The press of mill business drove the thought from his mind, and with it went Mt. Healthy's chance to become another famous watering place, like French Lick or Saratoga Springs."[239]

Figure 12.6. Water from the well on the mill property was found to have a high mineral content and medicinal properties.

There has been speculation that the medicinal quality of the water could have been a reason the people of Mount Healthy were spared from the cholera epidemic that swept the area in the 1840s. Even without a spa, the flat open area in front of the mill that led down to the creek was a popular spot for corn roasts and barbeques during the time the Hartmanns owned the mill.

Mill for Sale

*"**A FIRST-CLASS ROLLER** MILL, 6 room house and barn, one mile north of Mt. Healthy. Address Chas. Hartmann, Mt. Healthy, Ohio. C. D. &T. Traction Stop 86."* [240]

Twenty-seven years after he'd purchased the mill, Charles Hartmann was ready to retire (Figure 12.7). He was sixty-five years old, suffered from asthma, and had been ill with pneumonia several times.

As they prepared to retire, Charles and Amelia bought four lots on the corner of Elizabeth Street and Adams Road in Mount Healthy for five hundred dollars, and later sold two. They built their home there, at a cost of $4,580.59.[241]

Claude C. Groff and his wife, Carrie, purchased the mill, and once the sale was complete, the Hartmanns moved into their house on Adams Road. Charles died seven years later, during the influenza epidemic of 1918. He is buried in Saint Mary's Cemetery in Mount Healthy.

When Hilda married Leslie Fischvogt, the young couple stayed in the family home to care for Amelia until her death in 1946. Their children grew up in the house built by their

Figure 12.7. Norbert Charles Hartmann, 1911

grandfather, and after their daughter, Millie, married, she and her husband, Jim Hartman, continued the tradition and took up residence there to care for Millie's aged father.

I had the opportunity to meet with Millie Hartman in 2013, and enjoyed seeing the home that has been in her family for three generations. Millie's mother, Hilda, was thirteen years old when the family moved to Adams Road. Hilda had happy childhood memories of living near the mill, playing in the creek, and fishing from the covered bridge. She, her mother, and her sisters often paid visits to Orpha Rogers, who lived just up the road in the old Hill homestead.

Millie made sure I knew, for the record, that it was her grandmother, Amelia, who thought of the name Pride of the Valley.

She said the family was very close, and relied on their faith to get them through hard times. She also shared a family saying: "Wear your prayer book on the inside, not the outside," meaning to not be showy about one's faith, but to keep it in one's heart and truly live it.[242]

The Influenza Pandemic of 1918–1919

The influenza pandemic of 1918–1919 killed between twenty and forty million people worldwide. This was more people than were killed in combat in World War I, which saw sixteen million casualties, about two-thirds of which were in battle. The influenza pandemic has been cited as the most devastating epidemic in recorded world history. More people died of influenza in a single year than in four years of the Black Death from 1347 to 1351.[243]

The particularly virulent strain of the influenza virus struck young, healthy adults, with the death rates being particularly high for fifteen to thirty-four-year-olds, a segment of the population that was usually not affected by these types of infectious diseases. The disease struck viciously, and spread rapidly. Some died within a few hours of exhibiting symptoms.

In Cincinnati, citizens were warned on September 11, 1918, that the influenza had struck Boston and was working its way west. By October 5, there were an estimated four thousand cases in Cincinnati, and the mayor ordered all theaters, movie houses, schools, churches, saloons, and other public meeting places closed to slow the spread of the disease. Even with the quarantine, cases and casualty rates continued to climb, and by the end of October estimates were that Cincinnati might have had as many as twenty-five thousand cases of the disease. The city lifted all influenza-related restrictions on Armistice Day, November 11, and within a week, resurgence in cases of influenza was reported, especially among school children.

Around one thousand seven hundred people in Cincinnati died from the disease, a staggering sixty-four percent of those were people considered to be in the prime of their lives.[244]

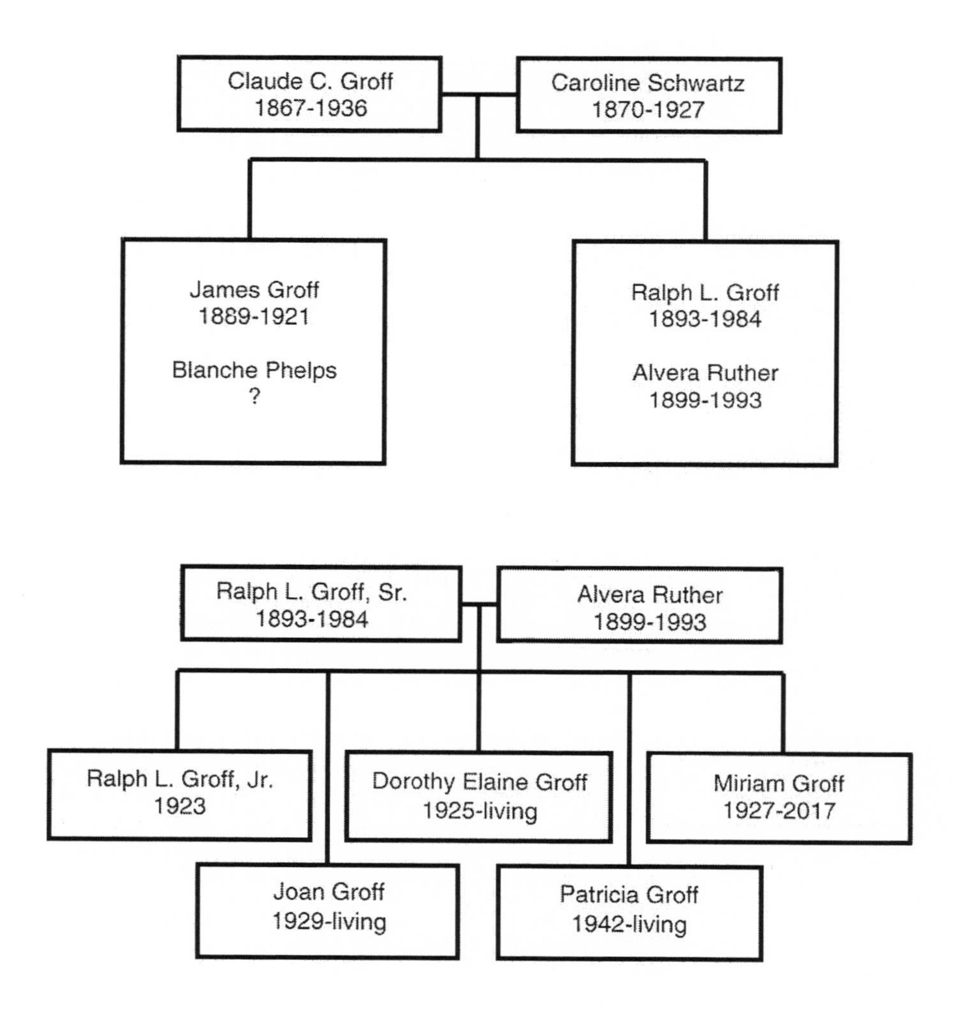
Claude C. Groff
1867-1936
Caroline Schwartz
1870-1927
James Groff
1889-1921
Blanche Phelps
?
Ralph L. Groff
1893-1984
Alvera Ruther
1899-1993
Ralph L. Groff, Sr.
1893-1984
Alvera Ruther
1899-1993
Ralph L. Groff, Jr.
1923
Dorothy Elaine Groff
1925-living
Miriam Groff
1927-2017
Joan Groff
1929-living
Patricia Groff
1942-living

13 C. C. and Ralph Groff — The End of the Independent Miller

Friendship and goodwill went into grinding grist for a neighbor.

– D. W. Garber, author of *Waterwheels and Millstones: A History of Ohio Gristmills and Milling*

Charles Hartmann placed an ad in the January 25, 1910, issue of *Grain and Farm Service Journal*, listing his mill and home for sale. In 1911, Claude C. Groff purchased the mill and the adjacent house, and took out a mortgage with Charles Hartmann for $8,500.00. The debt was to be paid in eight promissory notes, the first two for $2,000.00, the next six for $1,000.00, and the last for $500.00, payable annually. The deed also gave them the water rights to the dam on the Rogers' farmland, which had been conveyed to Charles Hartmann during the 1883 sale of the property. The Groffs paid the mortgage in full on July 16, 1915, four years ahead of schedule.[245]

C. C. Groff was born in Dayton, Ohio, on October 25, 1867. He attended Dayton Business College, and then learned the milling trade. He was the first owner of the mill to be college-educated, and was also the most widely traveled. He had first engaged in the milling business with his father in Traybines, Ohio, then followed his trade to mills in Lebanon, Ohio; Long Pines, Nebraska; Georgetown, Kentucky; Los Angeles, California; and finally to Mount Healthy in 1911.

The three years he and his family spent in Los Angeles was for the benefit the health of his wife, Carrie. It is both interesting and sad to note that the wives of Wilson Rogers, Charles Hartmann, and C. C. Groff all suffered from tuberculosis.

Help Wanted

We're able to get a glimpse of the day-to-day business through these want ads from 1913-1914, in a way not possible at any time previous in the mill's history:

> "Wanted: An engineer for small mill. One who has some experience in feed grinding and is willing and able to do other work about the mill. Steady position the year around. Must be a single man. In

replying state wages expected CC GROFF Mt. Healthy Ohio.[246]

Wanted — A reliable man, single preferred, to drive four-horse team; good home and board goes with this position; steady year round. References required. Address C. C. Groff, Mt. Healthy, Ohio, Stop 95.[247]

Secures Post in Ohio Mill

Editor American Miller — Please change my address from Chicago, Ill. to Mount Healthy, Ohio, as I have secured a position with C. C. Groff in the Mount Healthy Flour Mills through your paper. F. Pasikowski, Mount Healthy, Ohio." [248]

C. C. took his two sons, James and Ralph, into the business in 1914. The mill, which was then run by steam power, handled flour and corn products and did occasional custom grinding for local farmers.

Ralph, the younger son, was born December 17, 1893. During World War I, he enlisted in the U. S. Army on May 4, 1918, and served overseas for seven months as a private in the 303rd Battalion of the American Expeditionary Forces, Company A, Tank Corps (Figure 13.1). He was discharged with the rank of sergeant on April 9, 1919, and returned to the family business.

Figure 13.1. Ralph L. Groff in his uniform from World War I.

I had assumed James Groff and his wife, Blanche, were living in the house Charles Hartmann had built next to the mill in 1896, but information in the 1920 census entry makes it appear as if they were boarders in the Rogers home (Figure 13.2). It is also possible they lived in the mill house, as the Hartmann family had done. I haven't been able to resolve this this to my satisfaction. According to that same census, C. C., Carrie, and Ralph also lived on Mill Road, but their listing was two pages further along in the census records, so I wondered if they were living in another house, father away. Other sources say C. C. and Carrie occupied the house on the mill property." [249]

Covered Bridge Road now winds through the subdivision that was once Henry Rogers' farm, but the records from the early twentieth century refer to that road as Mill Road. In the present day, Mill Road terminates at its intersection with Miles Road, near the turnoff at the covered bridge.

Whatever the living situation in 1920, it changed drastically over the next few years. First, James Groff died on February 16, 1921, at the age of thirty-one. Within months of his brother's death, Ralph Groff took on a larger role in the family business. He and three other men incorporated the mill, and contracted to act as selling agent for a network of flourmills in Minnesota and Kansas.

> "September 1, 1921:
>
> Capitalized at $50,000 the CC Groff Milling Company has been incorporated at Cincinnati Ohio. The incorporators are Louis Weiland RL Groff EH Laufenberg and HJ Sohrer. The company has taken over the old plant of CC Groff & Son at Mount Healthy Ohio. RL Groff will supervise the mill and

◀ **Figure 13.2.** This charming image of Blanche Phelps Groff, James Groff's wife, shows that while many mechanical and technological improvements were a regular part of life in the early part of the twentieth century, some household chores hadn't changed much in recent centuries. I found this photo in a box of my own family's photos. The writing on the back says, "Blanch Groff, Wash Day."

the sales end will be under the management of FH and WA Laufenburg. The company will also act as selling agent for the Eagle Roller Mill Company New Ulm Minn and Newton Milling & Elevator Company of Newton Kan."[250]

Ralph married Alvera Ruther in 1922. The young couple moved into the house near the mill and lived there with Ralph's parents.[251] Their first child, Ralph L. Groff, Jr., was born in April, 1923, and lived only eight days.[252] Tragedy struck the Groff family yet again in 1924, when the frame house built by Charles Hartmann was destroyed by fire, taking with it many of the Groffs' possessions.[253] The North College Hill Volunteer Fire Department, which had been established in 1919,[254] answered the call for help in the department's Model T fire engine. According to the North College Hill Fire Department's website, the biggest fire the Model T fire engine ever answered was at the C. C. Groff home on Mill Road. The home was destroyed, but the fire department saved the mill by pumping water from the nearby creek.[255]

After that fire, the Groffs lived for a time in the mill, and then with friends, while they had two new houses built on the site, one for C. C. and Carrie and another for Ralph and his family.[256] C. C. took a loan of $4,500.00 from the Mount Healthy Savings and Commercial Bank on July 14, 1924, and repaid the loan, which was for "house erected on Mill lot, Mt. Healthy Mill Rd" by August 8, 1924.[257] Those homes, one a brick foursquare and the other a brick bungalow, are currently owned by Great Parks of Hamilton County, and have served as both office space and as homes rented out to park employees.

Ralph and Alvera's daughter, Dorothy Elaine, was born in 1925, and was followed by Miriam in 1927 and Joan in 1929. The mill's most popular product, high-ratio cake flour called "El-Mi-Jo," was named by using the first two letters of each of the girls' names (figures 13.3 and 13.4).

C. C. and Ralph had set a course to bring the mill into the modern age by incorporating the business and joining a network of flour brokers, and by making what became the last major improvement to the mill — converting it in 1923 from from steam power to a fifty-horsepower diesel engine.

C. C.'s wife, Caroline Schwartz Groff, died in December, 1927, following a short illness, at the age of fifty-seven. She was buried in Lebanon Cemetery in Lebanon, Ohio.

Like the previous owners of the mill, the Groff family was active in community, church, and civic groups. In 1929, Ralph was elected president of the Cincinnati Flour Club.[258]

In 1934, when C. C. could no longer operate the mill, Ralph took over operations. Ralph stated during a 1981 interview with Great Parks of Hamilton County employee Bob Lewis that his older brother, Jim, who had been the one who was extremely interested in the mill operations, had died at a young age, and Ralph "had to take over when his father was no longer able to operate the mill."[259]

Claude C. Groff died on April 28, 1936. The following month, the Groffs made a gift to their church in honor of Ralph's parents.

Figure 13.3. Ralph Groff's daughters Elaine, Miriam, and Joan.

Figure 13.4. Ralph Groff named several of the mill's products El-Mi-Jo for his three daughters.

"New Organ To Be Blessed

A new organ will be blessed and dedicated at 7:30 o'clock tomorrow night at St. Margaret Mary Church, North College Hill. A musical program will be presented by Henry Klosterman, church organist. Rev. B. J. Wellman, pastor, will conduct the service. The organ has been presented to the church by Mr. and Mrs. Ralph Groff as a memorial to Mr. Groff's parents, Mr. and Mrs. C. C. Groff."[260]

Ralph Groff continued to market Pride of the Valley Flour using the brand name chosen by Amelia Hartmann, and also developed other specialty products. These advertisements are our first look at how the owners had to adapt and change to meet changing times (Figure 13.5). The telephone numbers on the ads might seem quaint, or perhaps unrecognizable, to us!

In addition to expanding the business through advertising, marketing, and acting as a flour broker, Ralph Groff took meticulous care of the mill itself (Figure 13.6).

It appears that Ralph took his commitments — to the family business, his community, and his country — seriously. The mill ran three shifts of workers, twenty-four hours a day, to aid the war effort during World War II.[261]

The Allied victories in Europe and the Pacific brought an end to the war in 1945, but a crisis nearly as horrible as the war gripped dozens of nations during the winters of 1945 and

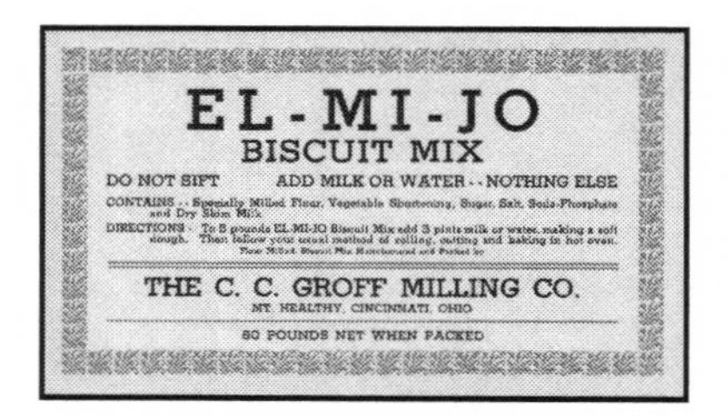

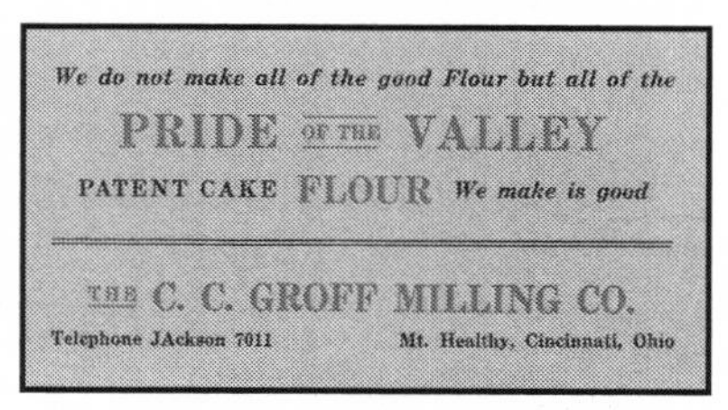

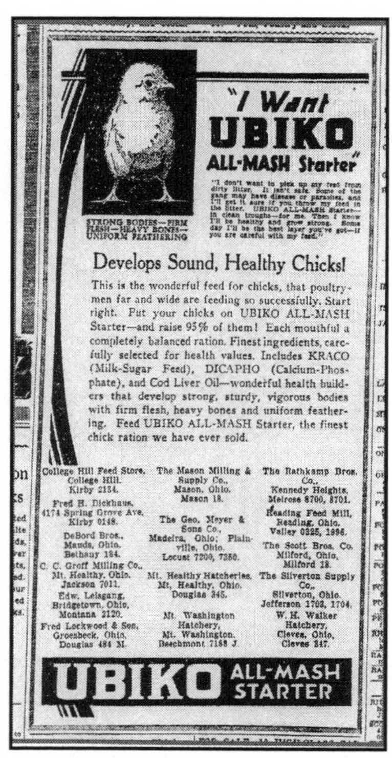

Figure 13.5. Ad cards for different Mount Healthy Flour Mill products, and an eye-catching design for their flour sacks.

Figure 13.6 Ralph Groff took pride in his business and kept the mill, by then a century old, in good repair. He had it repainted in 1948, just four years before it was closed for good.

1946. Millions of people, living on the edge of starvation, looked to the United States for help. The United Nations Relief and Rehabilitation Association (UNRRA) appealed for "mercy wheat" for the world's starving peoples.

President Truman called upon former president Herbert Hoover to assume the role of honorary chairman of the Famine Emergency Committee. Hoover assessed the need for relief in different countries overseas, while the Department of Agriculture set about buying up stores of wheat and corn.

The first shipments to Europe went out in late April, 1946.

"Mercy Grain Moves to Europe as Experts Predict Record Crop

Chicago, April 25 (UP) . . .UNRRA Director Fiorello H. La Guardia and Secretary of Agriculture Clinton Anderson headed into the heart of the grain area today to appeal for an increased flow of wheat from bins and storehouses to care for immediate needs until the winter wheat harvest in 20 to 40 days. They want farmers to free their supplies from the previous harvest so they can answer appeals of former president Herbert Hoover . . . for 1,100,100 bushels monthly overseas during April, May, June, and July. Those are the critical days in which wheat can avert famine.

The government has offered a 30-cent per bushel bonus on stored wheat and corn, hoping to raise 160,000,000 bushels of wheat and 50,000,000 bushels of corn." [262]

The appeal focused on humanitarian relief efforts, but there was trouble brewing. This program, founded with good intent, would soon see the mismanagement and the unintended results that can accompany large efforts of this kind.

"Seizure Needed Wheat Urged
OPA Declared Fascistic

Washington, D. C. April 27

Although the acute global food shortage may force this country to drastic action the agriculture department hesitated to approve any plan which might arouse the nation's farmers. The suggestion that the government take whatever wheat it needs was made by undersecretary of State Dean Acheson in a news conference late today.

He said the world food picture was bleak and would become more so. He said Americans were eating too much. The simple solution, he said, would be for the government to go to farms and mills and take whatever wheat and flour it needed." [263]

Within two weeks, newspaper articles criticizing the Mercy Wheat Program appeared on the front pages of national newspapers.

"Charges Government's Grain Bonus Creates Black Market

Hints of Rationing

Washington, D.C. May 4 (AP)

Senator Butler (R-Neb) accused the administration today of "plain, dumb, swivel-chair stupidity" for paying a bonus of 30 cents a bushel on wheat and corn for export while livestock and poultry producers are unable to buy feed.

The bonus was instituted to move grains off the farms for shipment to the hungry abroad. Butler asserted the lack of feed would force the "wholesale slaughter" of pigs, chicks, calves, and lambs.

Butler, a farmer and rancher, said that since only the government can legally pay the bonus, it has created "its own private black market in corn and wheat." Grain farmers face the decision whether to sell their corn for $1.50 to the government or $1.20 to a neighboring stockman.

By paying a bonus for corn the government forces hog raisers either to outbid the government in the black market or dispose of their animals." [264]

While the humanitarian intent of the Mercy Wheat Program was laudable, the ramifications of the methods by which it was handled soon became apparent.

"Wheat Farmers 'Strike' Refuse to Sell Grain

Washington, D.C. June 13

Wheat farmers in the great southwestern grain belt today were reportedly refusing to sell their newly harvested grain in a seller's strike against the government, dealing a new blow to hopes for an early end to the nationwide bread and flour shortage.

Flour mills are entirely dependent on new crop wheat to meet domestic bread-grain demands.

The small business committee is investigating the impact of famine relief food purchases on small businesses and charges that there has been mismanagement in famine relief administration."[265]

Small businesses were feeling the pinch, even as newsreel footage was celebrating the shipment of the first carloads of grain to starving people overseas. A relevant article appeared in the *Cincinnati Enquirer* in April, 1946.

"Flour Production Uncertain, Cincinnati Mill Officials Say

The outlook even for 75 per cent of normal activity in milling operations throughout the country in the next two months is not good, an official of one of the two Cincinnati flour companies actually doing milling in this vicinity, said yesterday.

What his concern, the C.C. Groff Milling Co., Mill Road, Mt. Healthy, makers of cake and pastry flour, will be able to grind in May is questionable, the official said. The mill has operated only part time this week. Any official of the other milling company said that mill might run one more day. Both establishments will remain open, however, to deliver processed products to their customers. The Groff concern has been rationing all accounts for some time, the official said.

Although the law now provides that millers must sell any grain they have on hand above a 21-day supply, at least 90 per cent of the nation's millers have not had even that much grain on hand."[266]

The *Enquirer*'s follow-up story in June, 1946, indicated the situation for the local millers had not improved.

"Wheat Shortage May Shut Down Two Flour Mills

Lack of sufficient wheat Tuesday slowed the wheels of Cincinnati's only two flour mills, which are threatened with a complete shutdown for the first time in a total of 121 years of milling operation.

R. L. Groff, operator of the C. C. Groff Milling Co., Mill Road, Mt. Healthy, said his mill, begun in 1911, had not been closed since that date except for mechanical repairs, but that the disappearing wheat supply might force that move.

Henry Nagel, operator of the Henry Nagel & Son mill, 2168 Spring Grove Avenue, founded in

> 1860, said a mere trickle of wheat was coming into his mill and, while future closing was a possibility, he believed he could keep the mill open.
>
> Both mills normally receive wheat from farmers in an area of from 50 to 75 miles outside Cincinnati. Their milled flour is supplied to bakeries, restaurants, and some hotels.
>
> Groff said the wheat supply began to diminish early in April when the Federal Government first offered a bonus to farmers as an inducement to sell their wheat."[267]

By the end of 1946, Hoover made his recommendation to President Truman.

> "... the United States should never again place the distribution of American food or relief into international control (except for rehabilitation of children). The experience with UNRRA should be enough. We should coordinate any action that we may take with other governments but not accept any joint control. This applies to all schemes for international control of American food or relief as that can only end in foreign control of American farmers' prices and production.
>
> ... it is time to end government charity (except as mentioned, to children). It is my opinion also that the time has come to end even government credits for any purpose except to prevent starvation — and even here we should require some percentage of their exports at some future date to be set aside to repay us."[268]

The Mercy Wheat Project of 1946 could have had a devastating effect on many small businesses like the Groff mill, as intent and results are often two very different things. Even though technology, transportation, and communication had all come a long way in the hundred years since Jediah and Henry had converted the original mill to grind flour, the mill was still a small enterprise and the steps taken to advance the international relief effort had a negative outcome for the local families who depended on the mill for their livelihood.

Though Ralph Groff was understandably frustrated by the unintended consequences of the Mercy Wheat Project, we know he was not an uncharitable man. Two years later, in 1948, he was listed as a participant in the Ohio Food Train, which collected donations to be distributed to the needy in Europe by Catholic Rural Life, Church World Service, and Lutheran World Relief working on a cooperative basis.[269]

Memories of the Mill

Dave Huser of the Mount Healthy Historical Society, who has been indispensible during the research phase of this book, has a personal connection to the Groff mill. His father, Duke Huser, was a delivery driver for the Groffs from 1945 until the mill closed in 1952. According to Dave, it was all cash-on-delivery in those days, and his dad collected the payments as he went on his daily route. When it was time to settle up with the bookkeeper at the end of each day, he was usually twenty cents short because he'd bought two cups of coffee, one in the morning and one in the afternoon, while out making his deliveries.

As part of my research for this book, I had the pleasure of meeting and chatting with Ralph Groff's four daughters — Elaine, Miriam, and Joan, for whom El-Mi-Jo flour was named, and their youngest sister, Patricia. I came away from the interview feeling as though I'd acquired four delightful new relatives (Figure 13.7). We talked about how I came to write my first book, *Fips, Bots, Doggeries, and More*, and then I asked for their childhood memories of the mill.

Elaine remembered that she loved to reach into the chute as wheat bran flowed past, and take out a handful to eat. Miriam recalled that playing in the wheat bin on the top floor of the mill was like being in a ball pit in a kids' amusement area today — except, she said, you had to make sure you didn't go down the chute! Joan confided that she used to climb up in the trucks parked outside at the loading dock and play in the wheat, amid cries of "You did not!" from her sisters.

"I did, too!" she insisted.

Joan also remembered walking up the hill to visit Orpha Rogers, who still lived in the homestead, and that she found the wood stove at the old house intriguing. Patricia, the youngest, said her dad named Patsy's Cookie Flour after her, but it didn't sell as well as El-Mi-Jo.

Figure 13.7. The four daughters of Ralph and Alvera Groff, (from left) Patricia, Joan, Miriam, and Elaine, at our interview in 2013.

They all remembered when it was time to fumigate the mill — how the workers would tape off all the windows and doors then let off cyanide gas inside the building. They would return forty-eight hours later, in gas masks, to get rid of the dead rats, mice, and weevils.

I had noticed that on Ralph Groff's World War One draft card, he had claimed exemption from being drafted, on the grounds that he was needed in the family business where "we mill 40,000 bushels of wheat yearly and dispose of same in sales." I asked his daughters if their father had hoped for a deferment, but they insisted that their dad had wanted to go to fight for his country, and was proud to have been a part of the tank corps.

The Groffs' ownership of the Mount Healthy Mill spanned forty-one years. Though the era of the small, independent miller was passing away, Ralph Groff had taken steps to insulate his business against obsolescence by operating as a flour broker in addition to milling and selling his own high-quality product.

When the Army Corps of Engineers appropriated the mill and the land on which it stood as part of the Mill Creek Flood Control Project in 1952, Ralph Groff reluctantly left behind the mill, which had been lovingly maintained and was in perfect working condition. He built a new warehouse at 4460 Chickering Avenue and moved his operations there (Figure 13.8). The new location was off Spring Grove Avenue, just a few blocks from Procter & Gamble's Ivorydale plant in Saint Bernard. Ralph continued in the milling business for another twenty-three years, and retired in 1975.

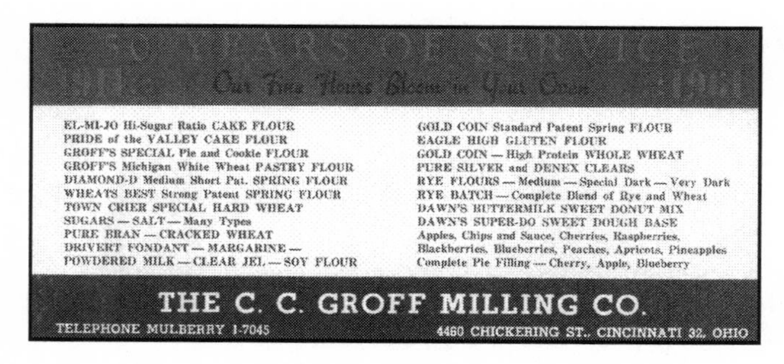

Figure 13.8. This ad card is dated 1961, and shows the new address on Chickering Avenue.

> "The MACHINERY and equipment of the former C. C. Groff Milling Co. are being offered for sale by the Corps of Engineers, Louisville, Ky., the Associated Press reported last night.
>
> The flour mill, at West Fork Mill Creek Reservoir, on Route 127, will be open for inspection until January 10.
>
> Sealed bids for the purchase of the machinery and equipment will be received until 11 a.m. in the office of the district engineer in Louisville."[270]

When Hamilton County Parks was exploring the possibility of restoring the mill, Bob Lewis, a park employee, interviewed Ralph Groff and went with him to the mill on May 29, 1981. Excerpts from his notes of the interview are transcribed below.

> "Met him at mill. Learned original wood house burned in 1922. Millrace believed to be east of Boone House in depression. We walked through the Mill — Mr. Groff recalled construction of diesel, one cylinder 50 hp engine on 35 cy concrete foundation. Then built enclosing shed at NW corner Mill Building.

Principal flour was "Pride of [the] Valley" a soft wheat general purpose. By-products were bran and middlins, which were sold as "Ubico" feed to cattle. Also produced El-Mi-Jo high-ratio cake flour, which was named after daughters Elaine, Miriam, and Joan.

[Groff] Is familiar with last location of mill equipment but was not primarily with production, heavily involved in sales and business management. Apparently built up large stocks of flour and reinforced mill floors accordingly. Was dismayed of mill's condition with effects of vandalism, equipment removal, and low maintenance since he sold to the Corps in 1952 in "perfect condition."

Said that stream bottom much higher in 1920s and they crossed West Fork of the Mill Creek with wagons and trucks just upstream of the barn-garage building. He took over Mill operation in 1934. His brother Jim, who was extremely interested in mill operations, died at an early age so that Ralph had to take over when his father was no longer able to operate mill.

Should talk again with Ralph Groff as he appears interested in our current efforts to determine feasibility and practicability of restoring mill to some significant point in its history."

I appreciate Bob's handwritten notes from this meeting because the unaltered and unedited comments gave an honest, even intimate look at Ralph Groff's emotions as he examined the mill site. He must have felt a great range of emotions when he observed the neglect and decay that had overcome the mill. He probably couldn't help but reflect on the early years, when his father and brother and he worked there together. The ghosts of the past, as they say, must have been stepping all over the heels of the present.

Section II

Place and Lore

Mill Creek — Essential, Dangerous, and Endangered

14

The waterway that invited the beginnings of industry in an unsettled country would later become known as the most endangered urban river in North America.[271]

Mill Creek is a 28.4 mile-long stream in southwest Ohio that originates in West Chester, in Butler County, and empties into the Ohio River along the south edge of Cincinnati (Figure 14.1). The watershed covers 166 square miles and thirty-seven local governments. Over one million people live and work in this watershed area on a daily basis.

The creek served as an extremely valuable and significant resource in Cincinnati's history. It was a hunting path for Native Americans, a power supply for pioneers' mills, a canal route for industries, and an escape route for fleeing slaves. In the early 1800s, settlers to the area named the stream Mill Creek, and soon mills, distilleries, slaughterhouses, and factories sprang up along its banks. In later years, international

Figure 14.1. Though Mill Creek has been part of the industrial side of the city, many parts of it appear unspoiled and picturesque.

Figure 14.2. Mill Creek has been used by farmers, mills, and other factories.

Figure 14.3. Mill Creek has also been used for recreation.

corporations such as Ford Motor Company, General Electric Aircraft Engines, and Procter & Gamble have operated alongside the Mill Creek[272] (Figure 14.2).

During the nineteenth and first half of the twentieth centuries Mill Creek spilled over its banks and wreaked havoc about every other year (Figure 14.3). The chief causes of this flooding were backwater spilling upstream when the Ohio River was in flood stage and storm runoff from the creek basin.[273]

All the owners of the Mount Healthy Mill had to deal with Mill Creek having either too much or not enough water. During the Hill/Rogers era, lack of sufficient water to power the mill was a constant issue. Hartmann and Groff used steam and diesel power to operate the mill, but these improvements couldn't keep back the floodwaters.

Flooding was only one of the problems associated with Mill Creek. The factories, slaughterhouses, distilleries, and mills that crowded the banks and built the city pulled water from the creek for their use. By the 1870s noticeable amounts of sewage and waste were being dumped back into the creek.

As area farmers sought to plant more crops, much of the flood basin was filled in and claimed as farmland, which further altered the creek's natural state (Figure 14.4).

By the 1890s, the Cincinnati Health Department pushed for a plan to do something about the stinking, polluted Mill Creek.[274]

Figure 14.4. Businesses have operated along the Mill Creek since the early 19th century. Runoff and other pollution led to it being labeled one of the most endangered urban waterways in America.

Floods were especially severe in 1832, 1847, 1853, 1907, 1913, 1918, and 1924, but the great flood of 1937, during which the Ohio River peaked at eighty feet, caused over seventeen million dollars in damages in Hamilton County, with losses in Mill Creek Valley accounting for about half those costs.[275]

The Red Cross called the flood of 1937 the nation's second-worst disaster of the twentieth century, behind World War I.[276]

Following the flood of 1937, the Army Corps of Engineers researched ways to curtail the floodwaters, and in 1939 began working on walling off the mouth of Mill Creek where it entered the Ohio River. After the barrier dam was completed in 1948, the Corps turned their energies toward flood control efforts on Mill Creek itself. A dam that would afford downstream protection and create a recreational reservoir for use by citizens was proposed. The project, which created the reservoir Winton Woods Lake, was authorized by Congress in the Flood Control Act and approved on July 24, 1946. The project was completed in 1952.[277]

The United States Corps of Engineers acquired the land on which the mill stood as part of the ongoing flood control project. This was the beginning of the end for the mill, which thereafter stood, abandoned and stripped of its equipment, on the banks of the creek that had supplied its power for most of its life.

The Corps considered another flood control project in 1981. Did this relate to local efforts to preserve the mill? It may have brought the mill back to people's attention, but the building was destroyed by arson in October of that year. The Corps project was abandoned in the early 1990s.[278]

Ironically, Mill Creek, which fed the mill and provided its source of power for nearly seventy of its one hundred thirty-year history, ultimately became the force that drove it to its destruction.

15

Sentinels from the Past at the Old Homestead

To cross over Mill Creek by way of the Jediah Hill covered bridge is to be surrounded by reminders of the past, even though an eclectic mix of homes built in the 1960s and 1970s now line one side of Covered Bridge Road. The barn built in the 1870s and the ruins of the mill are set back off the road, just to the left of the bridge. In front of the mill site stand the two homes built by the Groffs after fire consumed the Hartmann farmhouse (Figure 15.1).

The substantial frame house at the top of the hill was built by Jediah Hill sometime around 1830, and was home to four generations of the Hill and Rogers clan (Figure 15.2). Orpha Rogers Blake, Wilson's daughter, lived in the house until her death in 1964. But some people believe resident spirits have been there for a century and a half.

Whether you believe that the departed can remain tethered to earth, or you dismiss tales of haunted houses outright, I hope you'll enjoy these anecdotes about the ghostly occupants of the Jediah Hill homestead and draw your own conclusions about their origins and verity.

Kevin and Cindy Hardwick, the current owners of the house, graciously invited my family over for a tour the house in 2010, before *Fips, Bots, Doggeries and More* was published.

They spoke openly about the ghosts of Civil War soldiers that seem to have bivouacked in their home, and seem to take

Figure 15.1. The two houses were built by the Groffs.

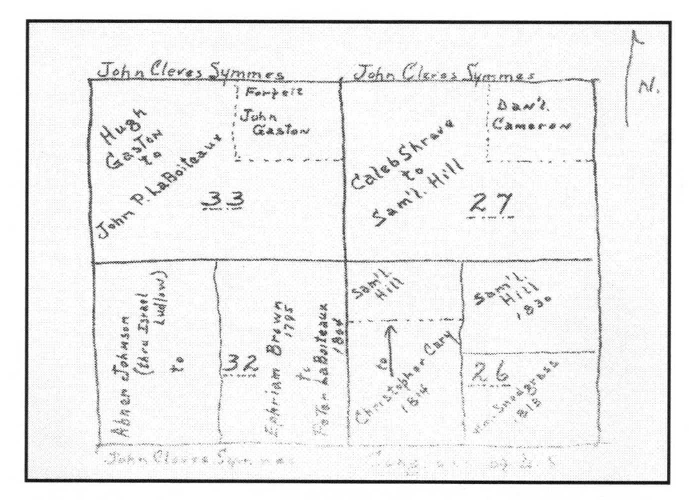

Figure 15.2. The Jediah Hill homestead, as it appeared in the 1990s.

the presence of their permanent houseguests in stride. According to Cindy, it's not unusual for their grandchildren to sleep over and come to breakfast the next morning, reporting, "Grandma! There were soldiers in the hall again last night!" Kevin says the spirits are friendly, and the supernatural activity seems to increase when they have family and friends over.

Once, while preparing for a family party, their daughter-in-law took some items upstairs, intending to store them in the back bedroom's closet. When she opened the closet door, she found the space occupied by a teenaged boy, dressed in the uniform of a Union soldier, complete with rifle and bayonet. She was so startled that she screamed and slammed the door. When she opened it again, no one was there.

I found this anecdote intriguing, though I wondered about the presence of the Civil War soldiers. I tried to reason it out — Wilson Rogers had served in the militia and the Ohio Volunteer Infantry, but he and all the men in his unit returned home unscathed.

I wondered if the specters could somehow be linked to Morgan's Raiders, who had passed through the neighborhood. This didn't fit either — Morgan's men were Confederate soldiers. There were no reported casualties in the area during Morgan's invasion on July 13, 1863. In fact, there was no evidence that any of the raiders had actually set foot in the house or the mill. What would prompt unearthly soldiers to bivouac at my family's house?

I mailed another round of questions to Aunt Nonie and Aunt Florence, but neither reported any knowledge of ghostly activity.

Fast forward to October, 2015. My parents and I visited Aunt Nonie, my grandmother's youngest sister, at the Masonic Retirement Home in Springfield, Ohio. We spent a pleasant day together and, of course, the conversation turned to my research for this book.

Aunt Nonie suggested we take her newspaper clipping of her grandfather Wilson's obituary down to the print shop where they could make photocopies. As she opened up a cabinet to retrieve the family Bible, she said matter-of-factly, "That's why the house is haunted, you know."

No. I did not know! Mom, Dad, and I looked at each other with expectantly raised eyebrows. One can never be sure what will cause long-forgotten memories to rise to the surface. I said, "Tell me more!"

Nonie stated that the Rogers family was very patriotic, and that during the Civil War, they fed Union troops and allowed them to camp on their land. Further, Wilson told the men in his unit that if they needed a place to rest on the way home, they were welcome to stop at his place.

I could imagine that any weary soldier who found himself in that bucolic setting, with rolling fields and the mill nestled down in the valley, would gladly stay for a while. Perhaps even a century or two.

Next I checked in with my cousin, Craig, and asked if his dad, Walter, who had been close to his grandfather, Pearl, and great-Aunt Orpha, might have heard stories about the ghosts.

Craig was quick to report back. "Walt had a very interesting story. He said that Orpha told him that during the Civil War, Henry and Wilson allowed Union troops to camp on their property and fed them. The troops respected them immensely, and were very loyal to them.

"Orpha told Walt that she knew exactly who the entities were because she recognized them. She said they are uniformed Union soldiers that are there to protect the homestead out of respect for Wilson. She saw them in the first floor hallway and in the pantry."

I now had plenty of evidence to conclude that the ghost stories were not just the product of, in the words of Ebeneezer Scrooge, "an undigested bit of beef, a blot of mustard, a crumb of cheese, a fragment of an underdone potato."

I'd barely had time to digest what I'd learned from Aunt Nonie and Craig when a woman named Carrie Kettell Parrett posted on the "Growing Up Mount Healthy" Facebook page. I recognized her maiden name, as I'd interviewed her mother, Carolyn, in 1991, during my preliminary research for *Fips, Bots, Doggeries, and More*.

Carrie's parents had purchased the homestead from Orpha's stepson George Blake after Orpha's death, subdivided it into two apartments, and rented it out.

During our chat on Facebook, Carrie relayed that as teens, she and her friends were convinced [the house] was haunted and spent nights in the cellar trying to conjure up spirits of Civil War soldiers. "Wrong era," she said, "but still."

She seemed excited to learn that other people had seen the ghosts. "I just know it was haunted! And I really don't believe in that sort of thing. . . . We spoke to them [spirits] with an Ouija board. No one ever told me it was Civil War. I thought we made that up!"

Carrie said her brother had moved into the front apartment, but returned to his parents' home after about a week because he didn't like seeing the ghosts.

I was excited that Carrie had been able to corroborate the others' experiences, but then she went them one better.

> "I have a super-weird memory. You'll love this: I practically lived in the large trees in the yard at the old house. There was one that I couldn't figure out

how to climb. I remember one day, a young man — he was young but he seemed old somehow — described to me exactly how to do it. I had never seen him before. He didn't live there. No one was home but me, and I was playing alone. I wondered who he was, but I did what he told me, and I successfully climbed into the tree. When I turned around to thank him, he was just gone. I think about that a lot. The memory of him teaching me to climb that tree has always been filled with a sort of magic, almost like a dream."

16 Renovations to the Covered Bridge

The Jediah Hill covered bridge was added to the National Register of Historic Places on March 28, 1973 (Figure 16.1). The bridge is of the Queen Post Truss design, in which two support posts are placed about a third of the way in on the framework, or truss, that can be used to support the sides or roof of a building (Figure 16.2). This design better supports a wider structure than a Kingpost design, which has one central support at the apex of the truss.[279]

The bridge is forty-four feet long with a twelve-foot deck width, and was constructed in 1850, shortly after the major expansion to the mill. The bridge was for use by the mill's customers, so it was in the best interest of the mill's owners to keep the bridge in good repair. After the mill was closed, the bridge was not as diligently maintained, and within two years, the deterioration was noticeable.

Figure 16.1. The Jediah Hill covered bridge was placed on the National Register of Historic Places in 1973. An Ohio historical marker commemorating Obed Hussey's 1835 test of the reaper in Jediah Hill's barley field will be installed nearby in September, 2017.

Figure 16.2. Interior view of the covered bridge, around 1978.

"County's Last Covered Bridge Appears Doomed

Cincinnati Enquirer, June 27, 1954

By Ed Seitz

If nature is allowed to run her course — as, unmolested, she has a habit of doing — the wooden superstructure of Hamilton County's last covered bridge will crumble into a picturesque stretch of the west fork of Mill Creek, near Hamilton Pike a few miles north of Mt. Healthy.

Why? Ralph Groff, who made it his personal concern to keep the bridge solid and functional for more than 20 years, no longer operates the mill to which the old bridge gave access.

The mill, in its own right a landmark of no mean proportion, was sold to the government two years ago when the West Fork Dam was nearing completion. (When the dam, now operative, is needed to protect the industrial valley from floods, waters from Mill Creek will back up to and inundate the old mill.)

The bridge, built in 1850, still serves about two dozen families who live on Covered Bridge Road. . . . After 104 years, it faithfully and safely bears the load of two steel I-beams between the automobiles and trucks, because Mr. Groff six years ago installed original stone pillars and wooden superstructure. But the sides and roof, which of course give the bridge its nostalgic charm and individual character, are rotting from lack of paint and bowing to the elements from lack of reinforcement.

Just a few yards downstream, after two more waterfalls, stood the great water wheel which supplied the power to grind the wheat to make the flour that went into the bread that was eaten in Greater Cincinnati before the turn of the century. The old bridge leans to the right and the roadway planking is in an advanced state of rot."

Not quite two years later, the rotting sides and roof collapsed from the weight of several snowfalls. The time for preventive maintenance on the bridge was past, and a major renovation was necessary.

"Citizens Hope To Save Covered Bridge

Cincinnati Enquirer, Sunday, February 12, 1956

The snows of January were too much for the weathered timbers of the bridge. A few weeks ago, the gently sloping roof collapsed under the weight of several snowfalls.

Local residents organized The Covered Bridge Committee, and "at the very first emergency meeting, $30 was subscribed to the cause. Since then an additional $5 has been received and the Springfield Township Board of Trustees have agreed to throw another $950 into the pot.

. . . the $950 would be sufficient to restore the floor of the bridge to the point where it would be safe for vehicles but that any additional funds for the sides and roof would have to come from other sources. The entire reconstruction job will cost approximately $3000, the committee believes, and it hopes that a public subscription will turn up to $2000 still needed to reach the goal [Figure 16.3]."

Figure 16.3. Orpha Rogers Blake stands beside the mill, after heavy snows in January 1956 caused the roof to cave in.

In the summer of 1977, plans to make the bridge "wider, higher, and safer" were underway.

"Old covered bridge to be renovated

Cincinnati Enquirer, 1977 article

By Lawrence Sussman

Plans . . . call for installing a concrete bridge deck to replace the current wooden one.

The bridge has a five-ton capacity, expected to be increased to 39 tons.

The $90,000 bridge renovation includes money for a Bailey Bridge that probably would be used during the nine-month repair job. The old bridge will be taken apart and its abutments and deck replaced.

The 125-year-old bridge is Hamilton County's only covered bridge, and residents want it to stay where it is. However, the New Burlington Fire Department, which serves the nearby Covered Bridge Acres subdivision, cannot get its new fire pumpers over the bridge. This led to the renovation plans [figures 16.4, 16.5, and 16.6]."

Though the 1977 article anticipated the bridge renovation would be completed sometime in 1978, there must have been delays. The temporary Bailey Bridge was still in place on October 24, 1981, which made it difficult for the New Burlington Fire Department to gain access to the scene the night the mill burned to the ground.

Figure 16.4. The bridge underwent a major renovation between 1978 and 1981, which was underway at the time of the fire that destroyed the mill.

Figure 16.5. Firemen used the temporary Bailey bridge to cross the creek and access the scene during the blaze.

Figure 16.6. The bridge as it appeared in the early 1990s.

Figure 16.7. The restored bridge decorated for the holidays in 2006.

17 Attempts to Preserve the Mill Come to Naught

Local informants contend that the original [mill] structure, in its entirety, exists within the altered building. Structural and mechanical modifications have taken place over the years, as part of the building's evolution from saw mill to a flour mill.

— from Army Corps of Engineers' Architectural Study of the mill

Currently, the covered bridge is well maintained, and able to bear the weight of local traffic (Figure 16.7).

By the late 1950s, the mill stood abandoned and stripped of its equipment on the edge of Mill Creek, which had been altered and somewhat tamed by the Army Corps of Engineers' West Fork flood control project.

Over the next two decades, the mill fell victim to vandals and trespassers. Neighborhood kids dared each other to go inside and smoke a cigarette or a joint. It sat back from the road, and was secluded enough to make it a popular place for amorous teenagers to park.

The mill stood on land purchased by the Army Corps of Engineers for construction and operation of the West Fork Flood Control project,[280] and when the project was finished, Great Parks of Hamilton County leased the land and the extant buildings. By the late 1970s, it was evident that something had to be done about the decaying structures (Figure 17.1).

There is plenty of conflicting information about what Great Parks of Hamilton County, then the Hamilton County Park District, intended to do about the mill. Any decision, whether it was to preserve the mill or tear it down, was bound to be costly. It is understandable if the parties responsible had changing opinions about what should be done.

Park officials requested the Corps of Engineers to assess the mill's structural integrity and historical significance. The Corps determined that it was eligible for the National Register of Historic Places, and began research and documentation in 1980. According to an undated interoffice memo from Jon Brady, the Corps was "not necessarily intent on disposing of the mill, and will not demolish it if [the park district] finds an adaptive use or are willing to preserve it." The Corps would

Figure 17.1. The last photo of the mill, taken before the fire in October, 1981.

not bear any expenses to restore the mill, but Brady was hopeful they might be able to cost-share with a grant for recreational facility funding.

According to the Historic Register nomination form, dated July, 1980, "The Park District has requested the Corps of Engineers to dispose of the buildings, while wishing to retain lease to the land."

Reporter Greg Loomis's July 1, 1981, article in the *Northern Hills Press* indicated that the mill would be torn down unless an acceptable way to save it was found. According to that article, local teenagers had repeatedly broken into and vandalized the building and its contents, and that fires had been started there.

When the mill was deemed eligible for placement on the National Register of Historic Places, the situation got even more complicated. Tearing it down would require review and approval by several federal agencies. The Park District reached out to the Miami Purchase Association and the Mount Healthy Historical Society, as well as private entities, hoping to interest them in restoring the mill. "We keep getting accused of wanting to tear it down," Brady said, "but we don't. We also don't want anyone getting killed in there."

The article reported that the Corps of Engineers awarded a $13,000 grant to University of Cincinnati professor of architecture J. William Rudd to do measured drawings and collect photographs of the mill, past and present.

The Historic American Buildings Survey/Historic American Engineering Record for the Mount Healthy Mill stated that the Corps of Engineers planned to demolish the Mount Healthy Mill. Under Executive Order 11593, mitigating documentation was undertaken in 1980 and 1981.

Those photographs, taken by the National Park Service and included in Appendix III of this book, give us a comprehensive look at both the interior and exterior of the mill as it existed at that time.

Dan Shaw, the current Director of Park Operations, was an intern with the Hamilton County Park District while the efforts to preserve the mill were underway. He constructed a scale model of the mill that could be taken apart so the different levels could be seen (Figure 17.2).

Figure 17.2. Dan Shaw constructed this scale model of the mill.

Who Burned Down the Mill?

I spoke with Jon Brady, then Director of Park Operations, about the mill fire when I was doing preliminary research for *Fips, Bots, Doggeries, and More* in 1990. He unhesitatingly called the fire arson, and stated, "A firebug burned it down. The arson squad did an investigation. The guy had set many fires, but there were no witnesses, and no proof."

I interviewed Glendale Fire Chief Kevin Hardwick, who was one of the first responders the night of the blaze, on June 4, 2013, at the Glendale Fire Station. The following is a transcript of his recollections.

> "I remember it being a Saturday night in late October, it was, it seemed unusually cold for the time, it was about 35, 40 degrees. When we originally got the call on the fire I want to say it's probably about nine o'clock at night, late evening. At the time I lived up on Mill Road, by Mill and Springdale...and we were dispatched for an unknown type of fire somewhere in the area of the covered bridge. It was reported by a Mount Healthy police officer. So we arrived on scene. At the time, the covered bridge was going through a rehab or a remodel. They had a temporary Bailey bridge set up next to it. They were also re-doing the water lines and the whole area by the mill was in disarray. They replaced a water line on the east side of the covered bridge. And there was a hydrant in what is currently my front yard. When we

arrived there was a lot of fire. It was reported that the cleaning crew at Carew Tower downtown, twelve miles away, could see the light in the sky [Figure 17.3].

We called for all the help in the world. We had obviously us, New Burlington, but we also called on Northern Hills, Colerain Township, North College Hill, and Forest Park. We had Greenhills Fire come in and cover the area around the park where Daley Road runs along the creek. All the embers were flying in that general direction, and we were getting little spot fires popping up in the park.

We figured, with the amount of BTUs being put out by the burning building, we would have needed somewhere in the area of 10,000 gallons of water a minute to get the fire under control. That was way beyond our capabilities. The first fire hydrant we tried, the one in our front yard, was full of rock, and immediately clogged up all our trucks. While some of us worked on getting everything cleaned out, others were pumping water down from Mill Road, almost to Saint Francis Seminary, which was the next good water source we had.

It seemed like the fire went on forever. We had issues going back to the first two houses on Covered Bridge Road. There were trenches dug in in the yard, and I can remember people going toward the fire and falling in those trenches — it was like

Figure 17.3. The enormous fire threatened the surrounding area. The light from the flames could be seen in downtown Cincinnati, twelve miles away.

a battlefield. Some of them had water in them and you didn't know how deep it was until you stepped or fell into it and then you might be four feet deep in water.

At one point, we had a problem with the barn near the mill — it's still there. That building was literally smoking on the outside from the radiant heat off the mill, so we had to put additional trucks down there to protect the barn [Figure 17.4].

Later in the fire — I have no idea what time — probably midnight, one o'clock, the building came down. I was on the back side of the building, close to the smoke stack, and I remember somebody yelling, "It's coming down!" so we turned and ran and as the building came down, it hit the back of me and pushed me out and I fell so fast that it landed on my leg and pulled my boot off, but I had enough momentum to get out. That was the closest I've ever come to being injured on a fire scene.

Figure 17.4. Firemen wet down the barn near the mill to keep it from catching fire.

One of the guys from Northern Hills was with me, and he did get hit pretty good by the flying debris. We took him to get checked out. One of our guys was standing in front of the big six-foot-wide barn door when the wall went out, and he was able to stand where the door was and the wall fell all around him.

I can remember that like it happened last night.

We were there well into the next day. The Sheriff came down and set off a controlled blast to bring down the chimney [Figure 17.5].

The next day the site was still too hot to get anything done, so we went back on Monday and started going through everything. We got down to where the millstone was in the lower level. We tried to reassemble anything we could, but the building was almost completely empty and cleaned out. There was nothing there, except we did find a metal garbage can lid. Our investigation found that someone had gone in and built the fire in that lid.

Figure 17.5. The building was a total loss.

I don't know if we can classify it as arson. We investigated two possible suspects and scenarios. First was a guy who had been up at the Lucky Lady bar and was walking home near the mill. It was possible he had gone into the building and set a fire in the metal garbage can lid to keep warm. We interviewed him a couple of times, but there wasn't enough evidence to prove that was the cause of the fire.

In the other case, there was talk of someone who lived on Covered Bridge Road that had heard about Hamilton County Parks planning to develop the site as a restaurant or gift shop or some other function, and this individual was not happy about the thought of increased traffic and activity in the neighborhood. The theory was he set fire to the mill to get rid of it.

The arson investigators talked to him a couple of times as well, and again, there was nothing we could pinpoint. So it really all just kind of evaporated away. There wasn't a clear-cut owner, so no one pushed for the investigation.

It was a long, busy night. I remember pulling out of my driveway and thinking, this is one of those nights when you're going to be gone a while. At the time I was a fairly young, new officer in the department. It was my first big fire, and one of the biggest fires I've ever seen to this day.

When you have a five-story building completely engulfed, it's a challenge.

That one picture was taken from what's now my front yard." [281]

Three days after the fire, on Tuesday, October 27, 1981, the *Cincinnati Enquirer* ran "Park Officials Ponder Life without Grist Mill," a story by Walt Schaefer.

"Springfield Twp. — Hamilton County Park District officials are mulling over where to go now since their dream of restoring the county's last water powered grist mill evaporated in flames during the weekend.

The five-story, 160-year-old structure burned down early Sunday, leaving only its stone foundation and a few charred beams.

Cause of the fire remains under investigation. The Specialized County Arson Task Force (SCAT) has been called to investigate, but that is standard procedure for all fires, said Kevin Hardwick, New Burlington Fire Department spokesman.

The mill, located on a creek across Covered Bridge Road near the intersection of Miles and Mill Rds. in the county park district, was built in 1820. Several additions were built on later.

Jon Brady, director of operations for the park district, said he had been trying to get funds to restore the building, which until several months ago was used for storage of vehicles in the winter and to paint signs and garbage barrels. He estimated the minimum value after restoration at $155,000, although other estimates exceeded $1 million.

Stu Welch, park director-administrator, said the park district may consider rebuilding the structure from drawings that were made of it, or, as a long shot, perhaps re-assembling a similar building from somewhere in the country.

A stumbling block, he conceded, is the probable cost of both options, and private funds likely would have to be tapped in an eventuality.

"How do you replace something that's the last of a dying breed?" Welch asked. "Would it be valid to try and restore the building? I don't know. We'll have to re-evaluate our direction."

Brady said the building had been emptied in the hope of getting aid to restore it.

The mill operated commercially into the early 1950s when it became part of the federal purchase for the flood-control basin of Winton Lake. It alone of 34 water-powered grist mills in the county escaped demolition."

An interoffice memo from Jon Brady, dated October 26, 1981, summed up his disappointment.

"Subject: The Groff Mill

The Groff Mill was destroyed by fire on Sunday morning, October 25, between 1-3 AM. All that remains is the stone and concrete foundation. The fire is presently being investigated by the arson squad to determine its cause and to explore the possibility of arson. The area should remain undisturbed and protected until the investigation is complete.

A large amount of effort has been made by maintenance and rangers to secure and patrol the mill. Several individual staff members also expended a great amount of time and study into exploring the mill and its future uses. I'm sorry that everyone's hard work has been for nothing. Even though this historic structure is gone, we can say we did try to protect and preserve it.

Presently, I am not sure what direction we will take. As our major emphasis was to preserve the existing building from the foundation up. The location just doesn't lend itself well for public use. We do want to eliminate any safety hazards. Perhaps the foundation can be filled in or moved — we'll explore the options and do what is best."

Minutes from a November 19, 1981, meeting at Hamilton County Parks Board read:

"Upon motion made and seconded the Board approved the staff recommendation to restore the mill site by collapsing about half of the stone walls into the foundation and filling with dirt. This was approved by the Corps of Engineers and is acceptable to Miami Purchase Association. The Board approved receiving bids on an emergency basis to demolish and restore the area as soon as possible for reasons of safety."

And finally, in a letter dated November 24, 1981, from Jon Brady to Ken Baker of the Colerain Township Police:

"Dear Ken:

Any fire or destruction of property by fire is most unfortunate, however the loss of a historic structure is especially saddening. The Groff Mill in Winton Woods stood for over 100 years only to be totally lost in less than one hour. The staff at the Hamilton County Park District and the citizens of our area feel a real loss due to this unnecessary and tragic fire, however we very much appreciate your efforts and the efforts of the investigative team in trying to determine its cause.

Thank you very much for your assistance.

Sincerely,
Jon Brady
Director, Park Operations"

I contacted Bob Mason, an employee of Great Parks of Hamilton County, for his recollections, and he responded to my email in July 2014.

"In the late 70's to early 80's Great Parks set out gathering information and devising a plan to restore the mill and use it for interpreting local history. I remember we sent a team to Clifton Mill near Yellow Springs, Ohio, to get some pointers on restoration and operation of a flour mill. We were just getting some local press coverage on our plans when the mill was "totally" destroyed by fire. I say "totally" destroyed because, although the wood structure was a total loss, the original stone mill built by Jediah Hill was exposed almost intact. It had been built upon and around over the decades and remained

> hidden and forgotten inside the wood building. Unfortunately, the heat of the fire had made the free-standing walls of the original mill unstable so the walls were knocked down and buried onsite for public safety.
>
> The fire department determined the fire started in the building's interior prompting the suspicion the fire was either deliberately set or resulted from an accidental ignition by an intruder. Perhaps a homeless person or a sightseer attracted by the press reports."

Mount Healthy Mayor James Wolf stated in a 2016 interview that, though he was too young to have any personal memories of the mill, it had always been part of his family's lore and legacy. His grandfather, Albert Wolf, a grandson of Charles Hartmann, was active in the efforts to preserve the mill, and was devastated by its loss.[282]

18 Renaissance

When I began writing this book, my two most motivating goals were to find out whether Jediah and Henry were the ones who converted the mill to grind flour, and to solve the mystery of who burned down the mill. I'm pleased that our research, which is detailed in the next chapter of this book, yielded proof that the mill was a saw-and-grinding operation during Jediah's lifetime but, after considerable effort and investigation, it became clear that I was never going to know exactly what happened to the mill on October 24, 1981.

Though I had to adjust my expectations and my goals as my research progressed, and though I was disappointed, I came to realize that what I learned about the mill and the families that owned it over those one hundred thirty years was more important than knowing who set the fire in 1981. *Pride of the Valley* turned out to be a story about people creating something of value, and how that value positively affected a community. When the mill was no longer valued, it was literally allowed to succumb to the flames of destruction.

Great Parks of Hamilton County has paid homage to the mill by renaming their public golf course on Sharon Road The Mill Course. During the 1993 renovation of the course, The Mill Race Clubhouse and Banquet Facility was constructed to look like a mill with a replica water wheel.[283] An informational display inside has preserved the history of the Mount Healthy Mill, complete with reproductions of Pride of the Valley flour sacks (Figure 18.1).

In 2009, the Hamilton County Park District obtained an original Mount Healthy Mill millstone from the yard of the "Covered Bridge House," which is located next to the former mill site. The millstone, along with three other millstones that had been acquired by the Park District, were cleaned and reconstructed by employee Bob Mason, and are now on display (figures 18.2, 18.3, and 18.4) as landscape features at the Mill Course and Mill Race Banquet Center.[284]

Of course I wish the mill still stood along the edge of Mill Creek. I wish school groups and history buffs and people who've never thought about how people got flour before there were grocery stores could "ooh" and "ahh" over the ingenuity, the workmanship, and the planning that went into creating the mill and making it a success.

Figure 18.1. The Mill Race Lodge and Conference Center at The Mill Course on Sharon Road is part of Great Parks of Hamilton County.

That's where the pride part comes in. *Pride of the Valley* isn't just the brand name for a product from a defunct business. It's a symbol of hard work and determination.

The town of Mount Healthy, like every small town in America, saw both prosperous and hard times in its two hundred-year history. People's lives were interwoven with major events that shaped our country's history.

The community saw times of polarity and discord, and other times when people worked together to survive, and to fight to right societal wrongs. It's a place where people hung

Figure 18.2. This millstone was buried in the yard of the Covered Bridge House near the mill site.

Figure 18.3. Buhr stones were often constructed of separate pieces held together in a metal frame.

Figure 18.4. Bob Mason of Hamilton County Great Parks with one of the millstones recovered from the property, now on display at the Mill Race Golf Course on Sharon Road.

on to what they had and worked hard to bring about something better. Descendants of the oldest settlers continue to live there, and refuse to give up on the town despite recent hard times, because they know what it has been in the past, and believe it is still a great place to live.

As the city prepares for its two hundredth anniversary, the Mount Healthy Renaissance has begun, and, like my pioneer ancestors, entrepreneurs are coming to the area to dream, try, succeed, and maybe sometimes fail. Citizens and civic leaders like Mayor James Wolf, a third-generation mayor of Mount Healthy, are emotionally invested in the area. Their roots go deep, and they've been joined by newcomers attracted by the city's affordable real estate and its proximity to amenities.

The Mount Healthy Renaissance is a nonprofit organization founded to encourage redevelopment in the area and educate the public about Mount Healthy's unique history. It advocates preserving historic buildings, like the Main Theatre at 7428 Hamilton Avenue. The Main has been closed since 1971, and, thanks to the efforts of local groups like the Mount Healthy Renaissance and Mount Healthy Historical Society, preservation and restoration of the historically significant structure are underway. The city bought the theater for $2,000.00 in 2015, and the structure has since been stabilized to prevent further deterioration. Even though the mill couldn't be saved, it's great to know that there is interest in preserving and repurposing the historic theater.

The Mount Healthy Renaissance also includes business owners like Bob and Betty Bollas of Fibonacci Brewing, a nanobrewery located at 1445 Compton Road. The Bollas chose to open their brewery in Mount Healthy because there were no other breweries near the community in which they lived.

May the pioneering and entrepreneurial spirit live on in Mount Healthy for another two hundred years.

19

The Mill's Evolution

Several clues helped Steve and me piece together a more complete history of the mill than has up to this time been available, and to establish an improved timeline of its evolution.

Clues

The Poem "The Old Covered Bridge"

Charles A. Hunt, author of "The Old Covered Bridge," was known as the Sage of North Perry Street. He was also Wilson Rogers's second cousin.[285] Born in 1869, Hunt was one month younger than Wilson's son, Walter. Perhaps the boys played together near the covered bridge (Figure 19.1).

Figure 19.1. The covered bridge, built by Jediah Hill in 1850, is the last remaining of its kind in Hamilton County.

There's an old covered bridge that stands today
Near the foot of a steep little hill
Where the roadway bends as it crosses the creek
That flows by an ancient mill.
On the farther slope, surrounded by trees,
Beyond where the millrace ran;
Still stands the cottage where I was born,
The home where my life began.

Pride of the Valley

I played, when a boy, by that Old Covered Bridge,
Played in the water and sand,
Imagined the bridge was a wonderful ship
And I the first in command.
Pretended the shadows were pirates, who crept
From the forest that stood on the ridge,
Seeking to capture my beautiful ship,
Capture that Old Covered Bridge.

Under the bridge I launched my fleet
With crews of courage and spunk,
But the foe crept in on the western flank
And my fleet, alas, was sunk.
That scene of battle is peaceful now,
No longer the cannon frown;
That Old Covered Bridge still marks the spot
Where a gallant fleet went down.

Over that bridge the farmers would come
Each bringing his grist of grain
When word went round that the miller would grind
As soon as we had a rain.
And the old gray miller would raise the gate
That ran in a moss-covered groove,
And the waiting waters would surge and push
Til the old mill-wheel would move.

But before the miller would grind the grain
Or ever the mill-stones roll,
He would carefully take from the farmer's grist
A measure of grain as a toll.
And I have learned as the years went by
In striving to reach some goal
That I must be willing, for favors received,
To pay to the Miller his toll.

And now I know from watching that mill,
Grinding so steady and slow,
What the poet meant when he wrote these lines
In the years of long ago:
"Though the mills of the gods grind slowly,
Yet they grind exceedingly small,
Though with patience He stands waiting,
With patience grinds He all."

That old mill-wheel has ceased to turn,
And the mill-race ceased to run;
The old mill stones, their faces worn,
Have vanished, their work now done.
The old gray miller now lies asleep
'neath the trees beyond the ridge;
But still the water flows over the ledge,
As it talks to the Old Covered Bridge.

It does not matter how long we live,
Nor matter where we may roam,
If we can live with the things we love,
We are living at "Home, Sweet Home."
May the Old Covered Bridge securely stand
As long as the waters run,
For I love most the things I saw
When my life had just begun.

When Hunt penned his verse "The Old Covered Bridge" in 1941, the water wheel was indeed stilled, the mill race filled in. Five generations of millers had gone to their eternal slumber "'neath the trees beyond the ridge." The mill was a modern factory, running on a diesel engine, and able to produce fifteen thousand pounds of flour a day.[286]

A close examination of "The Old Covered Bridge" yielded a number of details that support our thesis that Jediah and Henry expanded the mill almost immediately after they returned from their trip in 1838.

In the poem's fourth stanza, the miller raised the gate to allow the water to flow into the mill race, which Hunt called the "moss-covered groove." We can discern from this reference that the water wheel was used as a source of power in the early 1870s. Also, the line that the miller would grind "as soon as we had a rain" speaks to the problem of depending on rainfall to keep the creek high enough to generate the power to run the mill.

The Photograph

In their article about the Obed Hussey reaper, Rhode and Hite present as fact that the barn that stands near the ruins of the mill was Jediah Hill's original sawmill. Steve disagrees. He has found evidence of the mill race, and it didn't go anywhere near the barn. Further, Steve says, the barn's foundation was inadequate to absorb the movement and vibration of a working mill, and there was no evidence of a water wheel.

The Rhode and Hite article was already in print as of October 2014, and their statements were in direct conflict with our theory. We believed we were correct in our assertions, but how could we prove it?

A careful examination of a photograph of the mill that dated from the late 1860s, and was included in the Rhodes and Hite article, helped support our theory.

The barn that Rhode and Hite claim is the original sawmill, and that is extant on the site today, was absent from the 1860s photo. If the barn were the original sawmill, and built in the 1820s, it would have been visible in the left foreground of the photo, near the creek bank (Figure 19.2).

To the right of the mill, there was a structure that looked like a tiny cabin on stilts. This was the sluice gate, set at the point where the hand-dug mill race went underground as it headed toward the southern side of the mill. The sluice gate was used to control the flow of water to the wheel, which was located in the lower level of the sawmill.

The photograph also proved the business was operating as a flourmill in the 1860s. Steve asserts that there was

Figure 19.2. This stereoscope image, dated between the late 1860s-mid 1870s, is believed to be the earliest photograph of the mill. It shows the shed built over the sluice gate at the point where the mill race goes underground, and a small, whitewashed building between the house and the mill which may have been the Hill family's original cabin. (Image courtesy of Pat Brown of Brown Studios Photography, Hamilton, Ohio.)

absolutely no reason to build such a large structure for use as a sawmill, when a single-story structure with a cellar was adequate. A multiple-story building like the one in the photograph, however, was needed to accommodate the gears, millstones, elevators, bins, bolters, and other equipment used to mill flour.

Finally, the image showed what the mill looked like at the time from which Charles Hunt's boyhood memories derived.[287] It would continue to be useful in our efforts to understand what changes had been made at the mill, and when.

Good Bones

Steve and I were not the only ones who believed that the mill had always been in the same location, and that the original sawmill had been modified during the expansion of the structure to a grist mill. The Historic American Engineering Record of the Mount Healthy Mill, which was prepared by the National Park Service around 1980–1981, contains historical and descriptive data about the mill, including the statement that "local informants contend that the original structure, in its entirety, exists within the altered building"[288] (Figure 19.3).

Great Parks of Hamilton County employee Bob Mason had been able to observe the ruins of the mill the day after the fire, at a time when the outline of the original sawmill was evident. He made note that the "free-standing walls of the original mill" were unstable after the fire, and so the decision was made, in the interest of safety, to collapse the remains and bulldoze the area, to bury the ruins of the mill.[289]

Figure 19.3. The frame of the original sawmill is evident in this post-fire photo.

The Map

I could not have been more delighted when, in May, 2015, Steve emailed me a file containing an 1847 map of Hamilton County. I adore old maps, especially ones that have the landowners' names on them.

I hoped the map would help date the capital improvements to the mill, and my hopes soared when I saw Jediah's mill was marked with the notation G & S Mill. Aaron Lane's sawmill was downstream — also marked as G & S Mill. My elation vanished. Steve's research had established that Lane's mill had always and only been a sawmill.

Could the map be wrong? Was G & S Mill just a lazy way to mark mills without regard for their specialization?

Steve and I believed Jediah and Henry had completed the improvements long before — even decades before — the 1887 date often seen in local lore. But how to prove it? The answer was mere months away.

The Icepack That Connected It All

One Monday in February, 2016, after a particularly painful hour at the gym, I complained to my trainer, Lesley, about how much my knees hurt.

She regarded me sternly. Was I icing them after my workouts? Umm, no . . . I was usually eager to get going on my writing or dashing off on an errand after my time at the gym. Lesley admonished me to go home and ice for half an hour. No excuses. Chafing at the way my ex-dancer's knees were slowing me down, I went home and flopped on the sofa with two ice packs and my laptop.

On a whim, I decided to do a tangential search for John Lane, Jediah's friend and neighbor, on Ancestry.com. In the website's sidebar, which contained links to information about John Lane, was a census schedule I'd never heard of before: *Products of Industry in Springfield Township in the County of Hamilton, State of Ohio, during the Year ending June 1, 1850.* I clicked the link.

Aaron Lane's sawmill was highlighted on the schedule. It wasn't even John Lane's mill, but a tangent in that tangential search had led me to this particular schedule. I exclaimed aloud as I noticed the entry directly above Aaron Lane's sawmill was Jediah Hill's grist and sawmill!

There it was. The proof I'd been hoping to discover for more than two decades, and searching for actively for several months. I devoured the new information.

Jediah had reported investing $3,585 in the business that year. It was the second- highest capital investment in any business in the township, after the paper mill.

Joseph Corre, who is credited with starting the first mill in Springfield Township, reported a steam-powered sawing and corn-milling operation. According to the schedule, Aaron Lane's mill produced only lumber, and Jediah and Henry had the only wheat-grinding operation in the township in 1850.

Their raw material intake for the year included:
1,090 bushels of wheat valued at $900.00
3,150 bushels of corn valued at $1,417.00 and
150,000 feet of timber valued at $900.00.

The schedule confirmed that the mill used waterpower, and employed an average of two men at wages of thirty-two dollars per month.

The mill's output that year was:
41,000 pounds of flour, with a retail value of $1,029.00
3,542 bushels of cornmeal, with a retail value of $1,594.00, and
120,000 feet of lumber, with a retail value of $1,440.00.

Jediah reported:
Gross revenues of $4,063.00
Raw materials costing $3,217.00
Expenditures for wages of $384.00, and
Profit of $462.00.[290]

It is not known if the two men reported to be working in the mill were hired hands, or if they were Jediah and Henry themselves.

A More Complete Picture

After compiling our abundant information, Steve and I made an educated guess as to the mill's evolution through its one-hundred-thirty-year history. What follows is how we think it went.

Settlers were moving west to Ohio in large numbers, and there was a great demand for lumber for constructing homes and barns. Jediah, like all the other settlers, cleared his land of old-growth forest so he could plant crops. He saw the potential for a reliable income from the trees he felled, and commenced to build a mill with the help of his neighbor, John Lane.

The original mill was twenty-five feet wide by sixty feet long with a hand-cut creek stone foundation that was laid two feet thick and sixteen feet high. The building was constructed against a hillside with the entrance facing the hilltop level (figures 19.4 and 19.5). The water wheel, which was probably four feet wide and between twelve and fourteen feet tall, was inside the basement level, which helped keep it from freezing in the wintertime.

As water cascaded in from the race and over the wheel, it would take about two hundred gallons per minute to get the wheel turning, and four hundred gallons per minute to run the mill. The clean-out area below the wheel would have been between six and ten inches deep, with a floor that sloped toward the tailrace. The area under the wheel would quickly fill with water, and the tailrace would carry it away just as quickly to prevent friction loss from water pooling under the wheel (figures 19.6 and 19.7).

The water wheel turned gears that operated both the saw blade and a wrought iron carriage track, which moved the bucked[291] logs toward the saw as the blade ran with a

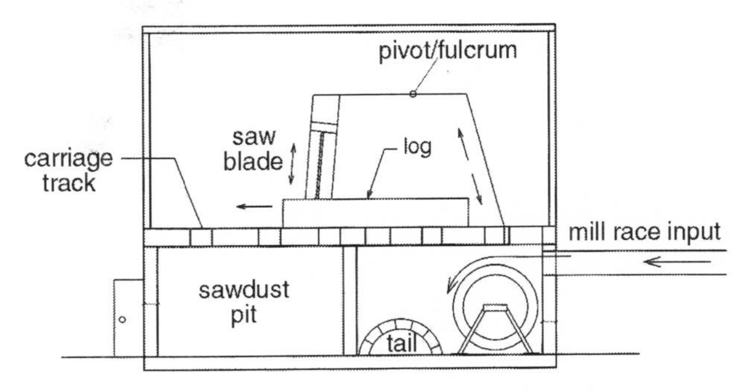

Figure 19.4. Cross-section of the water wheel and saw rooms of the original sawmill.

Figure 19.5. The water wheel in the original mill would have been very much like this one, at Newlin Grist Mill near Concordville, Pennsylvania.

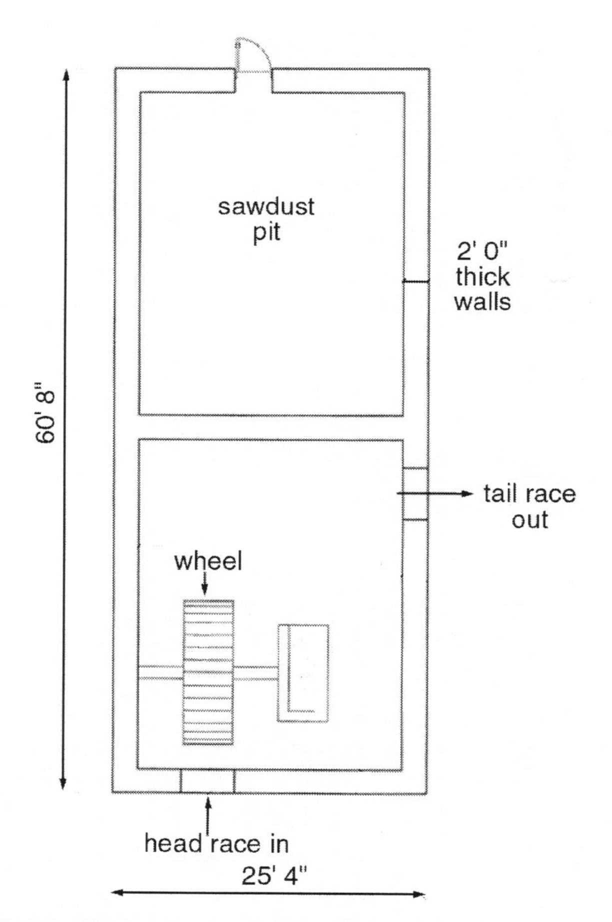

Figure 19.6. Layout and dimensions of the lower level of the original sawmill.

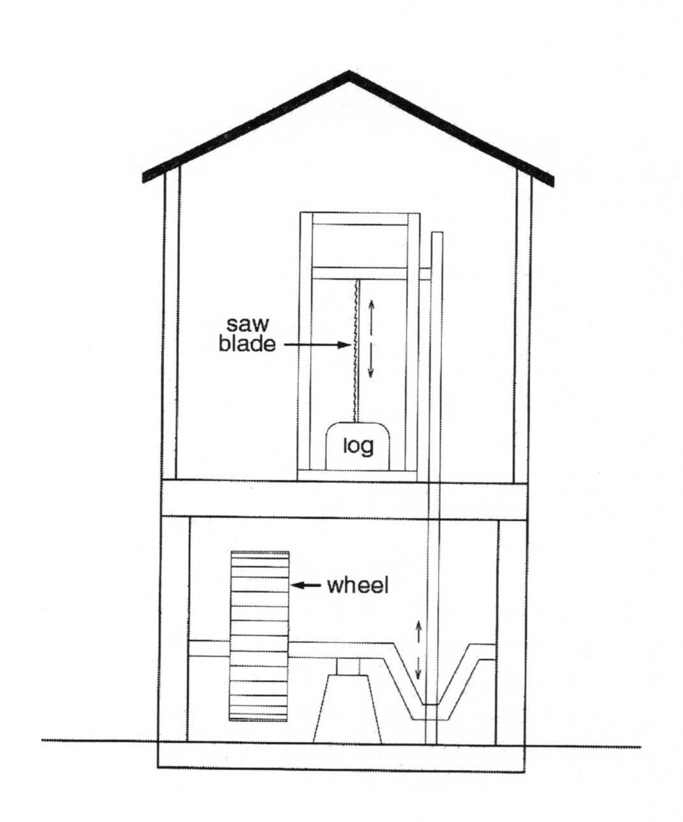

Figure 19.7. Cross-section view of the original sawmill.

twenty-eight inch up-and-down stroke. A separate room on the lower level was used to catch sawdust, which was sold for use as animal bedding.

The sawmill would have been constructed of hand-hewn beams, open on one side, with a rope winch powered by the same mechanism as the saw. The logs could be rolled off a wagon in front of the mill, then pulled in with a rope and rolled onto the carriage. The mill race, enclosed in an arched stone tunnel, ran directly under the area where the logs were dropped off.

The Saw

The saw in Jediah Hill's mill would have been a sash blade, so called because it resembled a square window (Figure 19.8). The water wheel connected to a crankshaft that pushed the saw blade up and down, and the sawyer controlled the speed of the saw by controlling the flow of water over the wheel.

Harnessing the Power Source

The millrace measured almost two hundred seventy yards from the dam to the mill's control gate at the drop above the water wheel (Figure 19.9). The last two hundred feet of the mill race was enclosed in a stone arch and a road built over it so wagons of logs could be driven up and unloaded, and then the sawn lumber carted out after finishing cuts had been done.

The tailrace, which carried water out of the mill and back to the creek, was one hundred fifty yards long, and was constructed of arched stone that ran for fifty-five yards underground, at an average depth of forty inches.

Figure 19.8. Most early nineteenth-century sawmills had sash blades, driven by a fulcrum, attached to gears that were turned by the force generated by water. The original Jediah Hill mill likely had one side open so logs could be rolled onto the carriage track.

More than one hundred twenty tons of stone would have been pulled from the creek bed, to be chisel-cut and-shaped by hand, and then used to line the headrace and tailrace.

Enough Water

The oldest dam, which was built upstream at the head gate for the mill race in Mill Creek during the construction of

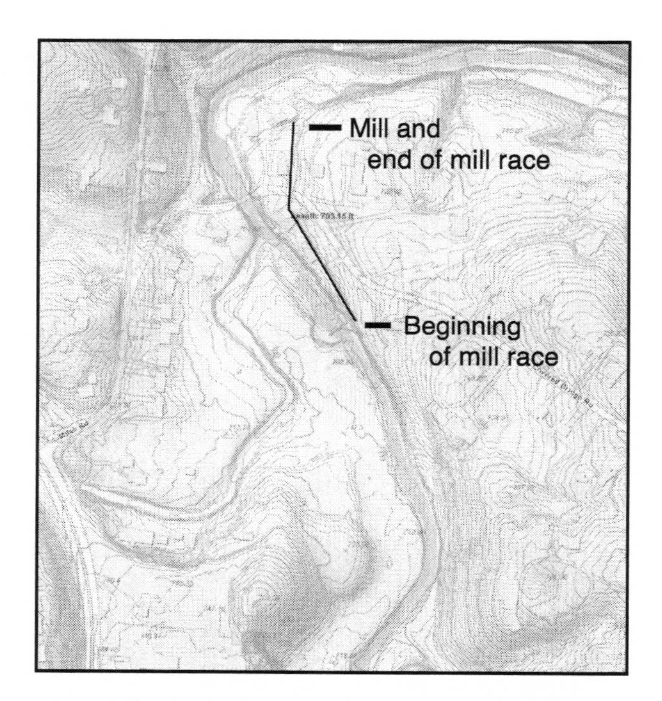

Figure 19.9. The mill race's path from Mill Creek to the mill was the length of nearly three football fields.

the original sawmill, would have been a timber crib dam, typical of the period. In this primitive type of dam, logs were laid in the stream bed and the gaps filled with rubble and soil. Creek stone was stacked at the edges, and more stone placed on top of the logs. Crib dams were vulnerable to washout and needed to be rebuilt every two to four years. The walls of the creek bank needed to be reinforced to hold the needed volume of water and prevent collapse, as the soil would become saturated and cave in.

The current dam on the site is eight feet high, but the earlier dams were likely ten feet high, based on the height of the abutment stones extant on the opposite bank.

On a ten-foot dam, the spillover point, plus six feet down, was useable for powering the mill. Once the water fell below the entrance to the head gate, the miller would send a hired hand out on horseback to open the gates on the other three dams that had been built in Mill Creek where it flowed through Jediah's land, and let in another charge of water. Once the water level rose to the top of the dam, the process would begin again.

During dry times in the summer when there was no water flow, no movement, and no ground water, water storage was a huge concern. Without it, there was no way to power the mill. Jediah and Henry may have built a replacement dam at the time of the improvements the mill. The new dam would have been laid of hand-cut stone and filled with rubble masonry, and angled in a V shape for strength and at least fifteen percent more capacity (Figure 19.10).

This dam was one hundred forty feet in length and ten feet high, and may have been rebuilt several times. It was used until 1900–1910. Dams of this type often washed out during spring freshets and experienced frost heaving, or damage to the masonry joints, from the freeze and thaw cycles of the seasons.

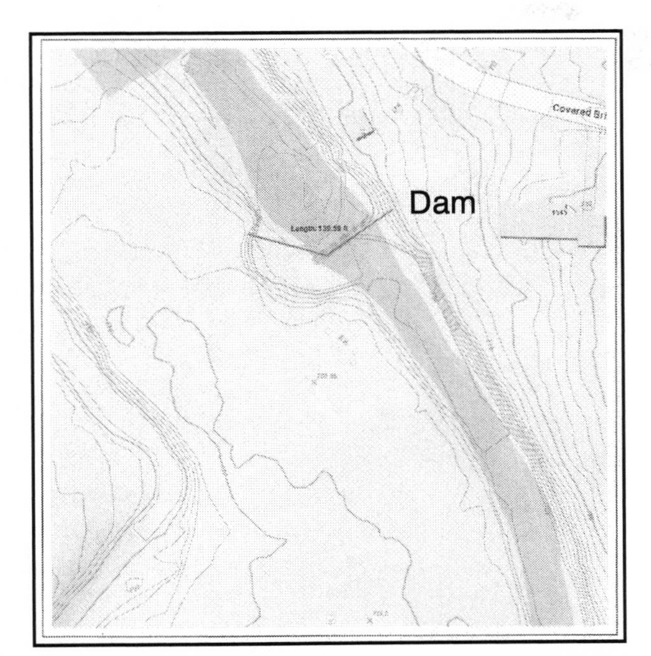

Figure 19.10. The first dams, located near the headgate for the mill race, were V- shaped.

Mill Creek is prone to very high water velocities due to rainfalls and spring melts. The water swell time is very fast, with six-foot swells and speed increase of nine to twelve knots. In more familiar terms, the water speed would have seen increases of ten to nearly fourteen miles per hour as the floodwaters surged. This type of rapid change in volume and velocity can be very destructive to dams (Figure 19.11).

The empty retaining wall pocket for the V-shaped dam and the headgate for the mill race are visible today. The retaining wall itself was removed when sewer pipes were laid in the creek bed in the twentieth century (figures 19.12 and 19.13).

Capital Improvements — 1839–1847

Steve and I have narrowed down the timeframe for the major capital improvements to an eight-year span, and we feel confident that we are correct.

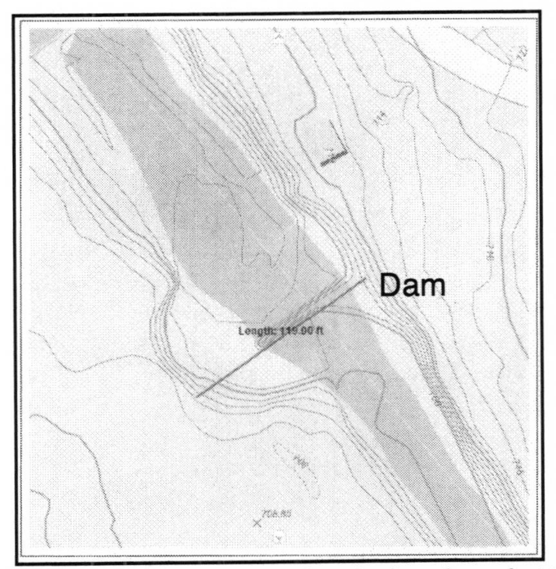

Figure 19.11. The most recent dam, built in the early 1900s, was completed after the switch to steam power.

Figure 19.12. Evidence of the side supports from the old V-shaped dam remains along the west side of the creek.

Figure 19.13. The headgate for the mill race can be seen on the east side of the creek near the dam.

We knew the 1838 trip was part of the planning stage, made before the expansion was to begin. Based on the speed at which the family traveled on the outbound portion of the journey, we estimated the return trip would take at least forty days. Therefore, we could infer that work on the expansion began sometime after they arrived at home in mid-to-late November, 1838.

The 1847 map, which had a notation about the grist and saw mill on Jediah's property, led us to the conclusion that the expansion was finished, or at least nearing completion, at the time the map was drawn (Figure 19.14).

Finally, the 1850 Industrial Schedule showed that the mill was operating at a profit, which we can offer as evidence that the mill had been operating in that capacity for at least a full year.

It would have been necessary to enlarge the original sawmill structure to accommodate the equipment used to process grain into flour. Each level in the new addition would have housed equipment that carried out one of the stages of the milling process.

In this diagram of a typical mill of the period (Figure 19.15), the lowest level houses the power source. The Mount Healthy Mill's water wheel was inside the structure, not

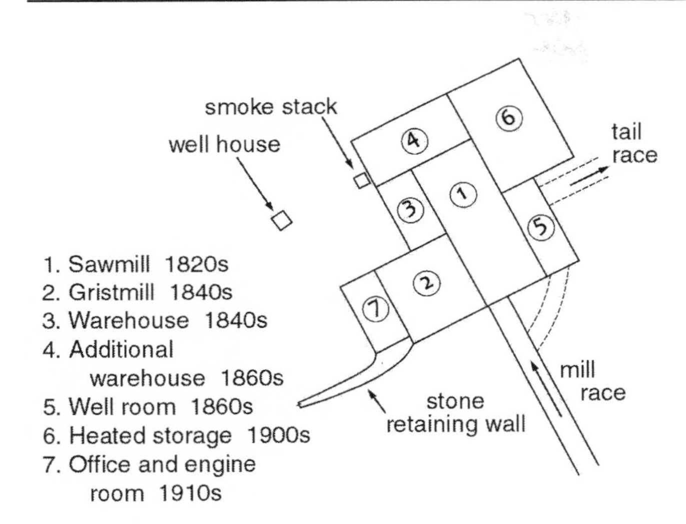

Figure 19.14. This plan shows the estimated dates of the establishment of, and additions to, the mill.

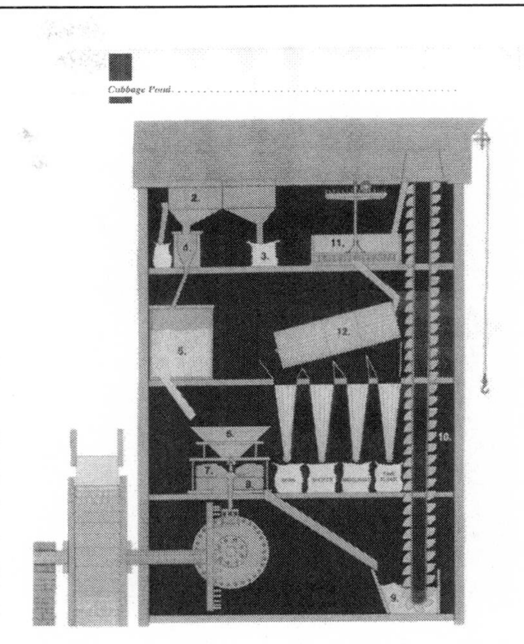

Figure 19.15. This is a cross-section of a typical water-powered flour mill. Three additional floors had to be added to the Mount Healthy Mill when it was expanded to produce flour. (Image courtesy of the Delaware Department of Transportation.)

outside as pictured in the diagram. The vertical water wheel was fastened to a horizontal drive shaft, and gears that changed the direction of the drive by 90 degrees carried the power to the saw blade. A great spur wheel may have been added to the water wheel, with additional gears that meshed with the stone nuts, which were the gears that drove the runner, or upper stones, that crushed and ground the grain.

The Grinding Process

The customer's grain sacks were first hoisted to the top floor of the mill with rope and pulley, and emptied into the dirty grain bin. The grain passed through a screen with a fan that whirled away dust, mold, and chaff which was collected in a sack. The grain then fell through the screen into the smutter which scoured off any remaining dirt. Next, the wheat would cascade down into a storage bin with a chute that opened to feed the grain into a wooden hopper, or funnel (Figure 19.16).

From the hopper, the grain could be measured into a hole at the center of the runner stone. As the runner stone turned, the outer hull of the grain was crushed between the grooves cut into the stones, and was ground against the flat parts of the stones, becoming finer as centrifugal force moved it out from between the stones.

Figure 19.16. Grain or corn was poured into the hopper suspended over the millstones, which are encased in this wooden frame.

The stationary bed stone would have been set in a frame, with the runner stone supported above on an iron spindle (Figure 19.17). The miller monitored and adjusted the distance between the stones while grinding was in progress, to assure the grain achieved the proper consistency and did not scorch from the friction between the stones.

The ground flour, moist and hot, was pushed out from between the stones into a tub built around the millstones. From there the flour was conducted down another chute to the floor below, only to be taken back to the top level in leather cups attached to a conveyor and deposited in the Hopper-Boy, a large tray where a rotating rake cooled and dried the mixture of flour, bran, middlins, and any foreign material that made it through the initial round of sifting at the dirty grain bin.

Lastly, the flour was sent down another chute into the bolter. There it was strained through a series of bolting cloths that progressed from tightly woven silk to a mesh screen. The bolter sifted out the finest flour, leaving the coarser outer layers of the wheat, known as the bran and middlins, which were then separated into different sacks to be sold as animal feed. If the miller added a portion of bran and middlins back into the white flour, it was considered whole-wheat flour.

It was a huge task to keep all the moving parts of a gristmill in perfect working order. The millstones required

Figure 19.17. The top, or runner stone, has been removed from this pair of millstones to show the grooves.

frequent maintenance, or dressing, a process in which furrows were cut with a steel pick to shape the grinding surface. A miller who dressed millstones could be identified by the discolored back of one hand, usually the left hand that held the pick. Minute particles of steel which flew from the pick during the dressing process would become embedded under the skin, thus giving the miller the look of a "steel-clad hand."[292]

Later, when roller mills threatened millstone grinding, many millers tried to improve the dressing of the millstone by increasing the number of furrows on their millstones. But even with enhanced furrows, the millstone still was slower at grinding than a roller mill.[293]

The chimney evident in photographs of the mill was constructed during these improvements, and was used to burn off the wood scraps from the sawmill and the chaff and waste from the grain milling processes.

Possible Improvements in the 1860s

There is evidence that Henry and Wilson decided to supplement the hydropower generated by the wooden overshot wheel with a more modern turbine.

Climatological records of the period indicate droughts in 1862, 1863, 1867, 1870, and 1874, about a third of the years of Henry's sole ownership of the mill.[294] There may also have been insufficient water to power the mill during years with average rainfall.

Evidence points to a second well room beside the original sawmill, constructed of stacked stone, but with a different method than the original foundation.[295] If a turbine had been added, it would have been an efficient source of auxiliary power. Turbines were smaller, metal water wheels with cupped blades that lay sideways beneath the surface of the water (Figure 19.18). Steve believes Wilson experimented with a Leffel turbine, which was available commercially from the Leffel Turbine Company of Springfield, Ohio, in 1862, and found there was still not enough constant water supply to feed the turbine.

The installation of a turbine was an expensive investment for any mill.[296]

Figure 19.18. During the 1860s improvements to the mill, Henry and Wilson may have added a turbine to supplement the water wheel. This example of a turbine from the period is on display at Indian Mill in Upper Sandusky, Ohio.

Charles Hartmann and the Switch to Steam

Charles Hartmann purchased the mill at a bargain price and ran it for fifteen years before undertaking a major modernization. In 1898, he made the conversion from hydro-powered water wheel to steam engine, and replaced the millstones with roller mills.[297]

During this renovation, the water wheel and the wall between the wheel pit and the dust room would have been removed to make room for the steam engine (Figure 19.19), and the wooden gears removed in favor of a belt-driven system.

The steam engine would have been a horizontal cylinder design, with one or two cylinders and a large flywheel

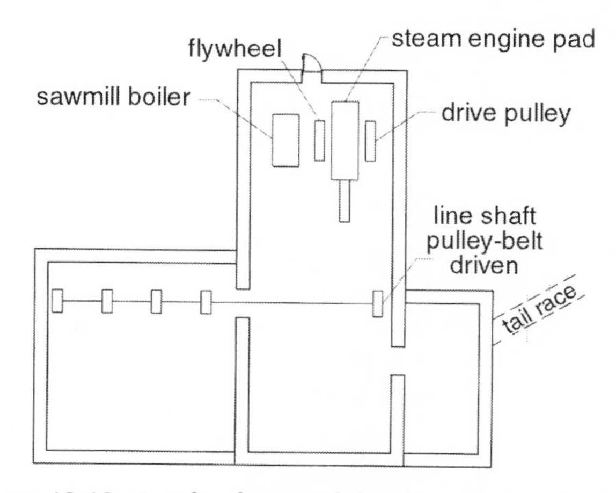

Figure 19.19. Overhead view of the alterations to the mill during its conversion to steam power.

(Figure 19.20). It would have been placed under the sawyer head of the upstairs saw.

The steam engine was located in the northwest corner of the building. The firebox that heated the water to make the steam burned scraps of lumber, and may have been set up outside the mill to reduce the risk of a fire in the mill.

At this time, the last warehouse and a loading dock may have been added on the northeast side of the building.

When roller mills first came into use, they were considered the first step in the milling process, used to break up the grain. A pair of millstones was then used to regrind the particles into flour. Eventually, as the roller mill technology improved, this practice became obsolete.[298]

Figure 19.20. The mill had a steam engine similar to this one at Baltic Mill in Baltic, Ohio.

The roller process uses a gradual reduction method, in which each step in grinding is performed with porcelain or steel rollers that rotate toward one another (Figure 19.21). In this process, the flour falls into a tray below the rollers after each grinding stage, and is bolted to separate the flour from the bran. Each successive pass takes the grain through finer rollers, working at different speeds.[299]

In 1898, Charles Hartmann contracted with the Orville-Simpson Company of Cincinnati to bring in the needed equipment. The millstones were removed and roller mills installed on the first level, the sifter/purifier in the basement, and the bolting and separating equipment occupied the upper floors.[300] The improvements allowed Hartmann to grind flour full-time, no longer a slave to the cycles of flood and drought that had plagued Jediah, Henry, and Wilson through their sixty years of ownership.

Hartmann kept the sawmill part of the enterprise for a few years after the conversion to steam, as evidenced by his son Charlie's data in the 1900 census, which lists his occupation as "sawyer in saw mill."[301] Sometime between 1900 and 1910, when the new system was up and running, Hartmann ceased to run the sawmill. When he advertised the mill for sale in 1910, he made sure potential buyers knew it was an updated roller mill.

Diesel Power

Claude C. Groff purchased the mill in 1911 and continued operations as a steam-powered roller mill. In 1917, Ralph Groff reported on his draft card that the mill had the capacity to process forty thousand bushels of wheat a year.[302] According to the National Association of Wheat Growers, one bushel of wheat yields forty-two pounds of white flour, or sixty pounds of whole-wheat flour.[303] If forty thousand bushes of wheat were all ground into white flour, the mill would have produced about 1,680,000 pounds of flour in a year.

Comparing the mill's output in 1917 with that of Jediah Hill's 1850 operation, we see that, with modern roller mill equipment and steam power, the mill's output increased by more than forty times.

It's important to bear in mind that we have a very small sample of information for our comparison, but we see that 1850, Jediah Hill reported taking in one thousand ninety bushels of wheat, and that his output was forty-one thousand pounds of flour. This averages to thirty-seven and six-tenths pounds of flour from a bushel of wheat, well below the sixty-two pounds of flour processed from a bushel of wheat with modern-day equipment.

After the Groffs converted the mill from steam to a fifty-horsepower diesel engine power source in 1923, the mill's output increased even more.

Ralph, who took over operations in 1934, was primarily involved in sales and business management. During his tenure, the business built up large stocks of flour and reinforced the floors accordingly (Figure 19.22). His mill served more than three hundred bakers, restaurants, hotels, and private customers in the Cincinnati area, and had the capacity to produce one-hundred-fifty hundred-pound sacks of flour daily. At that rate, and if the mill was in operation for two hundred fifty days per year, its annual output would have been 3,750,000 pounds of flour, more than ninety times the mill's output in 1850.

The shed on the far west side of the building was built to house the diesel engine.[304] This conversion eliminated the need for a fireman to run the steam engine, as the diesel engine was simple to start, and did not need to be monitored during the work day.

Ralph Groff oversaw operations at the mill until 1952, when he sold it to the Army Corps of Engineers "in perfect condition."[305] He then continued in business as a flour broker until his retirement in 1975. This notice, announcing the new tenant at the final location of the Groff milling business, appeared in the *Cincinnati Enquirer* on October 4 of that year:

> "The Goodyear Tire & Rubber Co. has taken a long term-lease on the property at 4660 Chickering St. The facility will be used for wholesale distribution. It was constructed for the C. C. Groff Milling Co. which has closed its doors after 62 years."[306]

Endnotes for Introduction and Sections I and II

[1] Thomas Longworth, *Longworth's American Almanac, New-York Register and City Directory for 1839* (New York, NY: 118 Nassau Street, 1839), p. 654. James H. Townsend, a dry goods merchant, is listed at 707 Greenwich Street. It is likely the family went there on an errand for a relative, Isaac Dukemenere, who owned a dry goods store in Fletcher, Ohio.

[2] US Department of the Interior National Register of Historic Places Inventory Nomination Form, accessed December 10, 2016; http://cdn.loc.gov/master/pnp/habshaer/oh/oh0400/oh0415/data/oh0415data.pdf.

[3] Carolyn Kettell, "Flours and a Handsome Homestead," *Cincinnati Enquirer Tristate Magazine,* October 2, 1988, p. 7.

[4] Ohio History Central online, "Land Ordinance of 1785," accessed December 10, 2016; http://www.ohiohistorycentral.org/w/Land_Ordinance_of_1785?rec=1472.

[5] Sue Korn Wilson and Kathleen Mulloy Tamarkin, *Images of America: Mt. Healthy* (Charleston, SC: Arcadia Publishing, 2008), p. 9.

[6] Ohio History Central online, "Land Ordinance of 1785" op. cit.

[7] Indian Country Today Media Network online, "The War of 1812 Could Have Been the War of Indian Independence," accessed December 10, 2016; http://indiancountrytodaymedianetwork.com/2012/06/18/war-1812-could-have-been-war-indian-independence-118851.

[8] Richard P. McCormick, "The 'Ordinance' of 1784?" *William and Mary Quarterly,* Volume 50 Issue 1, January 1993, pp. 112–22.

[9] C. Albert White, *A History of the Rectangular Survey System.* (U.S. Department of the Interior, Bureau of Land Management, 1983), p. 12.

[10] *One Square Mile, 1817-1992.* (Mount Healthy, OH: Mount Healthy Historical Society, 1992), p. 2.

[11] Ibid.

[12] Ohio, Hamilton County Deeds and Mortgages, Volume O, Page 218.

[13] United States Federal Census, Year: 1820; Census Place: *Springfield, Hamilton, Ohio*; Page: *325*; NARA Roll: *M33_87*; Image: *266.* Samuel Hill is found in Springfield Township, with one female in his household between the ages of sixteen and twenty-six, and two females older than ten and younger than sixteen.

[14] Find a Grave online, "John P. Laboyteaux," accessed December 13, 2016; http://www.findagrave.com/cgi-bin/fg.cgi?page=gr&GRid=15828895&ref=acom. John P. Laboyteaux's gravestone in the Laboyteaux-Cary Cemetery at the corner of Hamilton and Galbreath Roads in North College Hill, reads "Erected to the memory of John P. Laboyteaux, a native of New Jersey, who departed this life March 4, 1842 aged 67 years, 10 months and 13 days." Several family tree sources on Ancestry.com place John P. Laboyteaux's birthplace as Raritan, NJ.

[15] Wilson, *Images of America: Mt. Healthy*, op. cit., p. 7.

[16] Leonard Hill and Louise Hill, *Descendants of Paul Hill and Rachel Stout through Charles Hill and of Moses Edwards and Desire Meeker through Uzal Edwards*, unpublished manuscript, 1953, pp. 5-8.

[17] Wikipedia, "Land Ordinance of 1785," accessed December 5, 2016;

https://en.wikipedia.org/wiki/

Land_Ordinance_of_1785#knep02.

[18] United States Federal Census, Year: *1850*; Census Place: *Springfield, Hamilton, Ohio*; Roll: *M432_685*; Page: *114A*; Image: *242*.

[19] Kettell, "Flours and a Handsome Homestead," op. cit.

[20] D. W. Garber, *Waterwheels and Millstones: A History of Ohio Gristmills and Milling, Historic Ohio Buildings Series Volume 2*. (Columbus, OH: The Ohio Historical Society, 1970), p. 4.

[21] David M. Gold, *Eminent Domain and Economic Development: The Mill Acts and the Origins of Laissez-Faire Constitutionalism*, accessed December 13, 2016; http://direct.mises.org/sites/default/files/21_2_5.pdf, p. 105.

[22] Garber, *Waterwheels and Millstones...*, op. cit., p. 20.

[23] Ibid., p. 52.

[24] Tracy Lawson, *Fips, Bots, Doggeries, and More: Explorations of Henry Rogers' 1838 Journal of Travel from Southwestern Ohio to NewYork City*. (Granville, OH: McDonald & Woodward, 2012), pp. 54-55.

[25] Ohio, Hamilton County Deeds and Mortgages, Volume 2, p. 328.

[26] Ohio, Hamilton County Deeds and Mortgages, Volume 46, p. 278.

[27] RootsWeb's WorldConnect Project: *Family Tree of Christopher W. Lane*, accessed December 12, 2016; http://wc.rootsweb.ancestry.com/cgi-bin/igm.cgi?op=GET&db=clane&id=I580.

[28] Henry A. Ford, A. M., and Mrs. Kate B. Ford, *History of Hamilton County, Ohio, with Illustrations and Biographical Sketches*. (Cleveland, OH: L. A. Williams & Co., Publishers, 1881), p. 371.

[29] William A. Taylor, "Living Ohio Sons of Revolutionary Sires," *The Ohio Magazine*, Volume 2, (January 1906), p. 168.

[30] Wikipedia, "1st New Jersey Regiment," last modified September 17, 2016; http://en.wikipedia.org/wiki/1st_New_Jersey_Regiment.

[31] Taylor, "Living Ohio Sons of Revolutionary Sires," op. cit.

[32] *A roster of Revolutionary ancestors of the Indiana Daughters of the American Revolution : commemoration of the United States of America bicentennial, July 4, 1976*. (Evansville, IN: Unigraphic, 1976), p. 545.

[33] United States Federal Census, Year: *1850*; Census Place: *Springfield, Hamilton, Ohio*; Roll: *M432_685*; Page: *131B*; Image: *277.The* 1850 Federal Census records indicate Henry and Phoebe's sixth child, Nancy B. Rogers Brown, was born in Pennsylvania in 1802.

[34] Ancestry.com, "Elizabeth Rogers' Life Story," accessed December 10, 2016; http://person.ancestry.com/tree/37564294/person/28139043307/story.

[35] Ford, *History of Hamilton County...*, op. cit.

[36] Ibid.

[37] Ibid.

[38] Taylor, "Living Ohio Sons of Revolutionary Sires," op. cit.

[39] Ibid.

[40] United States Federal Census, 1830; Census Place: *Mount Pleasant, Hamilton, Ohio*; Series: *M19*; Roll: *132*; Page: *256*; Family History Library Film: *0337943*.

[41] Ancestry.com, "Elizabeth Rogers' Life Story," op.cit.

[42] Ancestry.com, "Jonathan Holden's Life Story," accessed December 6, 2016; http://person.ancestry.com/tree/50654576/person/13905742049/story.

[43] Ancestry.com, "Michael Burdge's Life Story," accessed December 6, 2016; http://person.ancestry.com/tree/18905547/person/731044557/story.

[44] Ancestry.com, "Zebulon Strong's Life Story," accessed December 6, 2016; http://person.ancesty.com/tree/18905547/731044544/story.

[45] Hamilton Avenue Road to Freedom online, "The Cary Family," accessed December 9, 2016;
http://hamiltonavenueroadtofreedom.org/?page_id=1106.

[46] Ancestry.com, "Cyrus Brown's Life Story," accessed December 13, 2016; http://person.ancestry.com/tree/32708013/person/18315107175/story.

[47] Ancestry.com, "Maria Rogers' Life Story," accessed December 6, 2016; http://person.ancestry.com/tree/38161460/person/1928128992/story.

[48] Ancestry.com, "Cyrus Brown's Life Story," op. cit.

[49] Revolutionary War Pension and Bounty-Land Warrant Application Files, 1800-1900, [database online]. Provo, Utah, Ancestry.com Operations, 2010. Henry Rogers Application Record #19692.

[50] Taylor, "Living Sons of Revolutionary Sires," op. cit.

[51] Cincinnati Friends Meeting online, Historical Archive: Newspaper Clippings, accessed December 6, 2016; http://cincinnatifriends.org/about/meetinghistory.html.

[52] Find A Grave online, "Henry Rogers," accessed December 13, 2016; http://www.findagrave.com/cgi-bin/fg.cgi?page=gr&GRid=94820422&ref=acom.

[53] Michael Gunn, "18 Revolutionary War Patriots finally honored at Wesleyan Cemetery" *Cincinnati.com*, October 19, 2015. http://local.cincinnati.com/share/story/225417.

[54] Lawson, *Fips, Bots, Doggeries, and More,* op. cit., p. 50.

[55] Ibid.

[56] Lindley Harlow, "Western Travel, 1800-1820,"The Mississippi Valley Historical Review Volume 6, Number 2 (September 1919), pp. 167-168.

[57] Railroad Museum of Pennsylvania online, "The Empire Builders" Curriculum Guide Grades 5-8," accessed February 15, 2016; http://www.rrmuseumpa.org/education/

Curriculum%20Guide%20-%20Middle%20Level.pdf, p. 3.

[58] Ibid.

[59] Scholastic Facts for Now online, "Overland Travel Around 1800," accessed February 17, 2016;

http://factsfornow.scholastic.com/article?product_id =nbk&type=0ta&uid=10676862&id=a2022130-h.

[60] Glenn Harper and Doug Smith, *The Traveler's Guide to the Historic National Road in Ohio.* (Columbus, OH: The Ohio Historical Society, 2005), p. 4.

[61] Wikipedia online, "Alexander Lord Stirling," last modified December 5, 2016; http://en.wikipedia.org/wiki/William_Alexander,_Lord_Stirling.

[62] B. C. Benson, "A Northern California Herd," *Holstein-Friesian World*, Volume 18, Issue 2, (December 10, 1921), p. 87.

[63] Digital History online, *Percentage of American Labor Force in Agriculture*, accessed April 27, 2016;

www.digitalhistory.uh.edu/disp_textbook_print.cfm?smtid=1.

[64] About.com online, "The Agricultural Revolution," accessed April 27, 2016; http://inventors.about.com/od/indrevolution/a/AgriculturalRev.htm.

[65] Old Sturbridge Village online, "Lesson Plans: Farm Family," accessed April 27, 2016; http://resources.osv.org/school/

lesson_plans/ShowLessons.php.

?PageID=R&LessonID=34&DocID=2037&UnitID=

[66] Digital History online, *Percentage of American Labor Force in Agriculture*, op. cit.

[67] Old Sturbridge Village online, op.cit.

[68] Ibid.

[69] Agriculture in the Classroom online, "Growing a Nation: The story of American Agriculture: Historical Timeline –Farm Machinery & Technology," accessed April 27, 2016; www.agclassroom.org/gan/timeline/farm_tech.htm.

[70] Robert T. Rhode and Leland Hite, "Obed Hussey And His Ohio Test of the First Successful Reaper." *Engineers & Engines Magazine*, October-November 2014, page 6. Passage cited with permission from the authors.

[71] Grace Estel Jones, *The Story of New Burlington, 1816-1922.* (Mount Healthy, OH: Hilltop Publishing Co., Printers, 1922), p. 14. This passage dates the construction of the stone cottage at 1841-42. If this were correct, the cottage was not built until after the Obed Hussey Reaper was perfected there in 1835. Other sources date the cottage at 1813.

[72] Rhode and Hite, "Obed Hussey And His Ohio Test of the First Successful Reaper," op. cit., p. 17.

[73] *Mechanic's Magazine*, Volume 3, Number 4, April 1834, p. 194.

[74] Ibid.

[75] Follett L. Greeno, ed., *Obed Hussey Who, of All Inventors, Made Bread Cheap* (Rochester, NY: Follett L. Greeno, 1912).

[76] Ibid.

[77] Ibid.

[78] Clark Lane's letter to the editor of the Butler County newspaper *Ohio Democrat*, dated March 13, 1890, is on file at the Lane Public Library, Cummins Room, Hamilton, Ohio.

[79] About.com online, "Cyrus McCormick, Inventor of the Mechanical Reaper," accessed December 13, 2016; http://

inventors.about.com/od/famousinventions/fl/Cyrus-McCormick-Inventor-of-the-Mechanical-Reaper.htm.

[80] Francis P. Weisenberger, *The History of the State of Ohio, Vol. III The Passing of the Frontier, 1825-1850.* Carl Wittke, ed. (Columbus, OH: Ohio State Archaeological and Historical Society, 1941), p. 63.

[81] Ronald C. White, Jr., *A. Lincoln: A Biography*. (New York, NY: Random House, 2009), pp. 212-213.

[82] Explore Pennsylvania History online, "Overview: Agriculture and Rural Life," accessed April 27, 2016; http://explorepahistory.com/story.php?storyId=1-9-4.

[83] Wikipedia, "Largest Cities in the United States by Population," last modified December 6, 2016; https://en.wikipedia.org/wiki/Largest_cities_in_the_United_States_by_population_by_decade#1830.

[84] New York Times online, Christopher Phillips, "The Breadbasket of the Union," accessed April 27, 2016; http://opinionator.blogs.nytimes.com/2012/08/08/the-bread-basket-of-the-union/.

[85] Ohio History Central online, "Agriculture and Farming in Ohio," accessed April 27, 2016; http://www.ohiohistorycentral.org/w/Agriculture_and_Farming_in_Ohio?rec=1579.

[86] Murray Rothbard, *A History of Money and Banking in the United States: The Colonial Era to World War II.* Edited with an Introduction by Joseph T. Salerno. (Auburn, AL: The Ludwig Von Mises Institute, 2002), pp. 95-99.

[87] Ibid.

[88] Wikipedia, "Second Bank of the United States," last modified November 28, 2016; https://en.wikipedia.org/wiki/Second_Bank_of_the_United_States.

[89] Michael F. Holt, *The Rise and Fall of the American Whig Party: Jacksonian Politics and the Onset of the Civil War* (New York, NY: Oxford University Press, 1999) p. 61.

[90] Wikipedia, "Second Bank of the United States," op. cit.

[91] John C. Hover, ed. *Memoirs of the Miami Valley in Three Volumes, Illustrated.* (Chicago, IL: Robert O. Law, Co., 1919), p. 197.

[92] Hill, *Descendants of Paul Hill and Rachel Stout...*, op. cit., p. 4.

[93] Lawson, *Fips, Bots, Doggeries, and More,* op. cit., p. 14.

[94] United States Federal Census, Year: *1850*; Census Place: *Springfield, Hamilton, Ohio*; Roll: *M432_685*; Page: *114A*; Image: *242.*

[95] United States Federal Census, Year 1820; Census Place: *Springfield, Hamilton, Ohio*; Page: *327*; NARA Roll: *M33_87*; Image: *268.*

[96] iupui.edu online, "A Brief History of the Abolitionist Movement," accessed December 13, 2016; http://americanabolitionist.liberalarts.iupui.edu.

[97] National Library of Australia online "Anti-Slavery Movement in the United States" accessed December 13, 2016; http://nla.gov.au/selected-library-collections/anti-slavery-movement-in-the-united-states.

[98] Hamilton Avenue Road to Freedom online, "Mt. Healthy and Anti-Slavery Society Conventions," accessed May 3, 2016; http://hamiltonavenueroadtofreedom.org/?page_id=793.

[99] Ibid.

[100] Ibid.

[101] Ibid.

[102] Hamilton Avenue Road to Freedom online, "Mt. Healthy and the Anti-Slavery Society Conventions," op. cit.

[103] Ford, *History of Hamilton County, Ohio*, op. cit. Henry's sisters and their spouses were listed in the profile of the family.

[104]Ancestry.com, *U.S., Appointments of U. S. Postmasters, 1832-1971* [database on-line]. Provo, UT, USA: Ancestry.com Operations, Inc., 2010. This collection was indexed by Ancestry World Archives Project contributors.

[105] Six Acres Bed and Breakfast online, "History of Six Acres," accessed June 5, 2016; http://www.sixacresbb.com/history.htm.

[106] Frank Woodbridge Cheney, *Subject: The Underground Railroad*. Cheney Family Manuscript Collection Ms 92876, Connecticut Historical Society, Hartford, Connecticut. 1901.

[107] Hamilton Avenue Road to Freedom online, "Other Mt. Healthy Abolitionists," accessed May 2, 2016; http://hamiltonavenueroadtofreedom.org/?page_id=829.

[108] *One Square Mile*, op. cit., p. 9.

[109] Mary Ellen Snodgrass, *The Underground Railroad: An Encyclopedia of People, Places, and Operations.* (London: Taylor & Francis, 2008), p. 550.

[110] Interview with Kristen Kitchen, owner of Six Acres Bed and Breakfast, February 2016.

[111] Six Acres Bed and Breakfast online, "History of Six Acres," accessed on April 24, 2016; http://www.sixacresbb.com/history.htm.

[112] Telephone interview with Butler County historian Jim Blount, April 2016.

[113] *One Square Mile*, op. cit., p. 19.

[114] *Reminiscental,* unpublished manuscript of autobiography of Clark Lane, on file at Butler County Historical Society, Hamilton, Ohio.

[115] Hamilton Avenue Road to Freedom online, "Mt. Healthy Christian Church," accessed May 3, 2016; http://hamiltonavenueroadtofreedom.org.

[116] D. Mylar Steffy, The Disciples Celebrationist online, "WWDBHS: What would David Burnet Have Said?" accessed December 16, 2016; https://disciplecelebrationist.com/2013/09/20/wwdbhs-what-would-david-burnet-have-said/.

[117] Ibid.

[118] Steffy, "WWDBHS: What Would David Burnet Have Said?" op. cit.

[119] Ibid.

[120] Ibid.

[121] Wikipedia, "Cincinnati Riots of 1836," last modified October 9, 2016;

https://en.wikipedia.org/wiki/Cincinnati_riots_of_1836.

[122] Patrick Young, Esq. Long Island Wins online: "Immigrant America on the Eve of the Civil War," accessed on April 17, 2016; http://www.longislandwins.com/news/detail/immigrant_america_on_the_eve_of_the_civil_war.

[123] Murat Halstead, *The Paddy's Run Papers;* (Unknown Binding, 1895) p. 202.

[124] Interview with Jim Blount, April 2016.

[125] Arthur J. Peterson, Chairman, Mount Healthy Sesquicentennial Celebration Committee. *Once Upon a Hilltop, Mount Healthy Area Sesquicentennial 1817-1967*, page 23.

126 Rootsweb online, "The Family Tree of Christopher W. Lane," op. cit.

127 Jones, *The Story of New Burlington*, op. cit., p. 8.

128 Ibid., p. 12.

129 Ibid., pp. 3-4.

130 Ibid., pp. 9-10.

131 Hover, ed. *Memoirs of the Miami Valley*, op. cit.

132 Brooke Julia, Ehow online, "The Duties of Township Trustees," accessed June 11, 2016; http://www.ehow.com/list_6569338_duties-township-trustees.html.

133 Jones, *The Story of New Burlington*, op. cit. p. 14.

134 Primitive Baptist online, "The Life of Wilson Thompson," accessed April 14, 2016; http://www.primitivebaptist.org/index.php?option=com_content&task=view&id=1405&Itemid=70.

135 Ibid.

136 Hamilton County Probate Court, Jediah Hill last will and testament, #5548, filed August 2, 1857, probated August 15, 1859. Box 13, Case 5547.

137 Ibid.

138 Frank Woodbridge Cheney. *Subject: The Underground Railroad*, op. cit., p. 8.

139 Wikipedia, "Ohio in the American Civil War," last modified May 27, 2016;
https://en.wikipedia.org/wiki/
Ohio_in_the_American_Civil_War.

140 Hover, ed. *Memoirs of the Miami Valley*, op. cit.

141 *Harper's Weekly*, Volume VII, Number 343, July 25, 1863.

142 *One Square Mile*, op. cit., p. 20.

143 Jones, *The Story of New Burlington*, op. cit., pp. 14-15.

144 Ford, *History of Hamilton County*, op. cit., p. 195.

145 *One Square Mile*, op. cit. p. 20.

146 Ibid.

147 Commissioners to Examine Claims Growing Out of the Morgan Raid, *Report of the Commissioners of Morgan Raid claims: to the Governor of the state of Ohio, December 15th, 1864*. (Columbus, OH: R. Nevins, State Printer, 1865) p. 109.

148 Wilson, *Images of America: Mount Healthy*, op. cit., p.12.

149 Ohio History Central online, *Cholera Epidemics*, accessed May 20, 2015; http://ohiohistorycentral.org/w/Cholera_Epidemics?rec=487.

150 *Titus' Atlas of Hamilton County, Ohio Entirely of Original Surveys* by R. R. Harrison, C. E. (Cincinnati, OH: C. O. Titus, Publishers, 1869).

151 D. J. Hebeler, *Mt. Healthy of Former Days: A History Compiled from Written and Oral Testimony*. Unpublished manuscript. Public Library of Cincinnati and Hamilton County Collection, pp. 3-4.

152 Nebraska Studies online, "Pioneer Children: School," accessed June 11, 2016; http://www.nebraskastudies.org/0500/frameset_reset.html?http://
www.nebraskastudies.org/0500/stories/0501_0207.html.

153 Ibid.

154 United States Federal Census, Year: *1860*; Census Place: *Springfield, Hamilton, Ohio*; Roll: *M653_979*; Page: *276*; Image: *180*; Family History Library Film: *803979*.

155 Primitive Baptist online, "Elder Wilson Thompson," accessed May 2, 2016; http://www.primitivebaptist.org/

index.php?option=com_content&task=view&id=1195&Itemid=55.

[156] New Madrid, Missouri online, "Strange Happenings During the Earthquakes," accessed December 1, 2016; http://www.new-madrid.mo.us/index.aspx?nid=132.

[157] Dr. David Stewart and Dr. Ray Knox, *The Earthquake America Forgot: 2000 Temblors in Five Months ...And It Will Happen Again!* (Marble Hill, MO: Gutenberg-Richter Publications, 1995) p. 249.

[158] *Official Roster of the Soldiers of the State of Ohio in the War of the Rebellion, 1861-1866, Vol. VIII*, Compiled Under the Direction of the Roster Commission. (Akron, OH: Werner Co., 1886-1895) pp. 671-2.

[159] Mabel Watkins Mayer, "Into the Breach: Civil War Letters of Wallace W. Chadwick," *The Ohio State Archaeological and Historical Quarterly*, Volume LII, Number 2, (April-June 1943), pp. 158-180.

[160] Hill, *Descendants of Paul Hill and Rachel Stout...*, op. cit., p. 50.

[161] Hover, ed. *Memoirs of the Miami Valley*, op. cit.

[162] Wilson Rogers obituary, unknown newspaper. Author's collection.

[163] Ancestry.com, "Amos Williamson Life Story," accessed June 11, 2016; http://person.ancestry.com/tree/69618910/person/30199920480/story. Amos' parents were Jacob Williamson and Sarah Hoagland, who married in Hunterdon County, New Jersey 1795.

[164] Hill, *Descendants of Paul Hill and Rachel Stout...* op. cit, p. 50.

[165] United States Federal Census, Year: *1870*; Census Place: *Springfield, Hamilton, Ohio*; Roll: *M593_1208*; Page: *652A*; Image: *371713*; Family History Library Film: *552707*.

[166] W. H. Alexander. *A Climatalogical History of Ohio, The Ohio State University Bulletin Vol. 28, No. 3, November 10, 1923*. (Columbus, OH: The Engineering Experiment Station of The Ohio State University. Printed by Ohio State Reformatory, Mansfield, 1924), pp. 308-309.

[167] The Social Welfare History Project online, "The Long Depression (1873-1878)" accessed June 11, 2016; http://www.socialwelfarehistory.com/eras/civil-war-reconstruction/the-long-depression/.

[168] Wittke, ed., and Eugene H. Rosebloom, *The History of the State of Ohio: The Civil War Era 1850-1873*. (Columbus, OH: Ohio State Archaeological Society 1944), pp. 7-10.

[169] Ibid.

[170] Ibid.

[171] Ohio, Hamilton County Deeds and Mortgages, Volume 465, pp. 51-53.

[172] United States Federal Census, Year: *1880*; Census Place: *Denver, Arapahoe, Colorado*; Roll: *88*; Family History Film: *1254088*; Page: *351A*; Enumeration District: *017*; Image: *0523*.

[173] Wilson Rogers obituary, op. cit.

[174] Chana Gazit, director, *The Forgotten Plague: Tuberculosis in America*, accessed April 12, 2016; WGBH Educational Foundation, 2015.

[175] Ibid.

[176] Ibid.

[177] Ashley M. Wilsey, "Half in Love With Easeful Death: Tuberculosis in Literature," Humanities.Paper 11 (2012), accessed May 31, 2016; http://commons.pacificu.edu/cgi/viewcontent.cgi?article=1010&context=cashu.

[178] Ibid.

[179] Gazit, *The Forgotten Plague: Tuberculosis in America,* op. cit.

[180] Jennifer Wirth, AllDay.com online, "In the Victorian Era Tuberculosis Actually Inspired These Beauty Trends," accessed June 4, 2016; http://allday.com/post/8457-in-the-victorian-era-tuberculosis-actually-inspired-these-beauty-trends/?exp=3&utm_source=ADS&utm_medium=FBO&utm_campaign=HIP.

[181] Ibid.

[182] Wikipedia, "Tuberculosis Treatment in Colorado Springs," accessed June 5, 2016, https://en.wikipedia.org/wiki/Tuberculosis_treatment_in_Colorado_Springs.

[183] Gazit, *The Forgotten Plague: Tuberculosis in America*, op. cit.

[184] *Tuberculosis History in Hamilton County* online, accessed June 5, 2016;
https://www.youtube.com/watch?v=Q8K34DSzaD8.

[185] Gazit, *The Forgotten Plague: Tuberculosis in America*, op. cit.

[186] *American Miller and Processor* Magazine, Volume 50 Issues 1-6, (1922).

[187] Hover, ed. *Memoirs of the Miami Valley*, op. cit.

[188] Stanley McClure, *The descendants of William Sater (1793-1849), Crosby township, Hamilton County, Ohio.* (Cincinnati, OH: S.W. McClure, Publisher, 1986).

[189] Hill, *Descendants of Paul Hill and Rachel Stout...* p. 53. William B. Hill's parents were Jediah's cousin Benjamin Hill and Eliza's sister Phoebe Hendrickson.

[190] Stanley McClure, *The descendants of William Sater...*, op.cit.

[191] Ibid.

[192] Ibid.

[193] United States Federal Census, Year: *1870*; Census Place: *Reily, Butler, Ohio*; Roll: *M593_1177*; Page: *419A*; Image: *431497*; Family History Library Film: *552676*.

[194] *The descendants of William Sater*, op. cit.

[195] Ibid.

[196] Melissa J. Homestead, *American Women Authors and Literary Property, 1822-1869. (New York, NY: Cambridge University Press 2005) p. 29.*

[197] Amanda Vickery, "Golden Age to Golden Spheres? A Review of the Categories and Chronology of English Women's History." *The Historical Journal.* 36.1 (June 1993) pp. 383-414.

[198] History Cooperative online, "The History of Divorce Law in the USA," accessed June 11, 2106; http://historycooperative.org/the-history-of-divorce-law-in-the-usa/.

[199] Gibson County (Indiana) Circuit Court, civil order book D, August 1829–February 1838, p 113.

[200] Ibid.

[201] U.S. Department of the Census, Marriage and divorce 1867-1906. (Ann Arbor, MI: University of Michigan Library, 1908) p. 299.

[202] Wikipedia, "Timeline of Women's Legal Rights" last modified June 10, 2016; https://en.wikipedia.org/wiki/Timeline_of_women%27s_legal_rights_(other_than_voting).

[203] Victorian Women: The Gender of Oppression online, "Historical Analysis: Women as "The Sex" During the Victorian Era," accessed June 7, 2016; http://webpage.pace.edu/nreagin/tempmotherhood/fall2003/3/HisPage.html.

[204] Ibid.

[205] US Department of Health, Education, and Welfare, *100 Years of Marriage and Divorce Statistics, 1867-1967*, accessed June 5, 2016; https://www.cdc.gov/nchs/data/series/sr_21/sr21_024.pdf, p. 9.

[206] Snopes.com, "1872 Rules For Teachers," accessed December 6, 2016; http://snopes.com/language/document/1872rule.asp.

[207] The Homeroom online, E. Graham Alston, Inspector General of Schools. "*The Government Gazette*, May 28, 1870," accessed December 16, 2106;

https://www2.viu.ca/homeroom/content/topics/statutes/rules70.htm.

[208] United States Federal Census, Year: *1880*; Census Place: *Okeana, Butler, Ohio*; Roll: *996*; Family History Film: *1254996*; Page: *182A*; Enumeration District: *025*; Image: *0735.*

[209] Hover, ed. *Memoirs of the Miami Valley*, op. cit.

[210] Find a Grave online, "Walter H. Rogers," accessed June 11, 2016;http://www.findagrave.com/cgi-bin/fg.cgi?page=gr&GRid=95374750&ref=acom.

[211] Hill, *Descendants of Paul Hill and Rachel Stout...* op. cit., p. 49.

[212] Death notice in unknown newspaper. Author's collection.

[213] Ancestry.com, US, Social Security Applications and Claims Index, 1936-2007 [database online]. Provo, UT, USA: Ancestry.com Operations, Inc. 2015.

[214] United States Federal Census, Year: *1900*; Census Place: *Springfield, Hamilton, Ohio*; Roll: *1283*; Page: *14A*; Enumeration District: *0318*; FHL microfilm: *1241283.*

[215] Ibid.

[216] United States Federal Census, Year: *1910*; Census Place: *Springfield, Hamilton, Ohio*; Roll: *T624_1196*; Page: *5B*; Enumeration District: *0344*; FHL microfilm: *1375209.*

[217] Ibid.

[218] Death Certificate, Nancy Gwaltney Rogers. Author's collection.

[219] United States Federal Census, Year: *1920*; Census Place: *Springfield, Hamilton, Ohio*; Roll: *T625_1396*; Page: *5B*; Enumeration District: *511*; Image: *725.*

[220] Hover, ed. *Memoirs of the Miami Valley*, op. cit.

[221] Death Certificate, Wilson Thompson Rogers. Author's collection.

[222] The German Americans online, "Why Germans Left Home," accessed April 19, 2016; http://maxkade.iupui.edu/adams/chap2.html.

[223] Ibid.

[224] Translation of Journal of Norbert Charles Hartmann, unpublished. Author's collection.

[225] Ohio, Hamilton County Deeds and Mortgages, Deed Book 557, pp. 434-436.

[226] Hartmann journal, op.cit.

[227] June 2013 interview with Millie Hartman, granddaughter of Charles Hartmann.

[228] Ibid.

[229] Ibid.

[230] United States Federal Census, Year: *1900*; Census Place: *Springfield, Hamilton, Ohio*; Roll: *1283*; Page: *14A*; Enumeration

District: *0318*; FHL microfilm: *1241283*.

[231] *Twentieth Annual Report of the Department of Inspection ofWorkshops, Factories, and Public Buildings to the Governor of the State of Ohio for the year 1903*. (Springfield, OH: The Springfield Publishing Company State Printers, 1904).

[232] United States Federal Census, Year: *1910*; Census Place: *Springfield, Hamilton, Ohio*; Roll: *T624_1196*; Page: *5B*; Enumeration District: *0344*; FHL microfilm: *1375209*.

[233] Ibid.

[234] Wikipedia, "Beer Garden," accessed April 15, 2015; http://en.wikipedia.org/wiki/Beer_garden.

[235] *Cincinnati Star,* Wednesday, May 13, 1925.

[236] Unpublished Hartmann family history. Tom Jesionowski collection.

[237] Terrace Park Historical Society online, "The History of Avoca Park," accessed June 11, 2016; http://tphistoricalsociety.org/the-history-of-avoca-park/.

[238] *One Square Mile*, op. cit., p. 45.

[239] *Grain and Farm Service Centers Magaine*, Volume 24, January 25, 1910.

[240] Unpublished Hartmann family history, op. cit.

[241] June 2013 Interview with Millie Hartman.

[242] Regional History from the National Archives online, "The Influenza Epidemic of 1918," accessed May 16, 2015; http://www.archives.gov/exhibits/influenza/epidemic.

[243] University of Michigan Library online, "Cincinnati, Ohio and the 1918-1919 Influenza Epidemic," accessed May 15, 2015; http://www.influenzaarchive.org/cities/city-cincinnati.html#.

[244] Ibid.

[245] *American Miller and Processor, Volume 41,* January 1, 1913.

[246] *Cincinnati Enquirer*, May 31, 1913.

[247] *American Miller and Processor,* Volume 42, June 1, 1914.

[248] Kettell, "Flours and a Handsome Homestead," op. cit.

[249] *American Miller and Processor,* Volume 49, Issues 7-12.

[250] Kettell, "Flours and a Handsome Homestead," op. cit.

[251] Find a Grave online, "Ralph Louis Groff, Jr." accessed December 6, 2016; http://www.findagrave.com/cgi-bin/fg.cgi?page=gr&GRid=59250926&ref=acom.

[252] Kettell, "Flours and a Handsome Homestead," op. cit.

[253] Cincinnati Fire Apparatus Resource online, "North College Hill Fire Department," accessed June 11, 2106; http://www.cincyfireapparatus.com/northcollegehill.html.

[254] *Cincinnati Enquirer*, July 14, 1969, p. 26.

[255] Kettell, "Flours and a Handsome Homestead," op. cit.

[256] Those homes, one a brick foursquare and the other a brick bungalow, are currently owned by Great Parks of Hamilton County, and have served as both office space and as homes rented out to park employees.

[257] *Cincinnati Enquirer*, December 8, 1929, p. 42.

[258] May 1981 interview: Hamilton County Parks employee Bob Lewis with Ralph Groff.

[259] *Cincinnati Enquirer*, May 23, 1936, p. 26.

[260] June 2013 interview with Ralph Groff's daughters.

[261] The *Stanford Daily*, Vol. 109, Issue 43, April 26, 1946.

[262] Madera *(CA)* Tribune, *Number 50, April 27, 1946.*

[263] *Chicago Tribune*, May 5, 1946, p. 7.

[264] Philip Warden, *Chicago Tribune*, June 13, 1946, p. 6.

[265] *Cincinnati Enquirer*, April 25, 1946, p. 11.

[266] *Cincinnati Enquirer*, June 4, 1946.

[267] Hoover & Truman, a presidential friendship online, "Part II: Feeding the World," accessed June 11, 2016; http://www.trumanlibrary.org/hoover/world.htm.

[268] *Cincinnati Enquirer*, August 7, 1948, p. 9.

[269] *Cincinnati Enquirer*, December 14, 1955, p. 36.

[270] Robin Corathers, "Life Blooming in Mill Creek," *Cincinnati.com* online, accessed December 13, 2016; http://www.cincinnati.com/story/opinion/contributors/2015/03/03/life-blooming-mill-creek/24325975/.

[271] Mill Creek Valley Conservancy District online, accessed December 14, 2016; https://sites.google.com/site/millcreekvcd/.

[272] Cincinnati Museum online, "Waterproofing the Mill Creek Flood Plain," accessed December 14, 2016; http://library.cincymuseum.org/topics/f/files/1937flood/wat-015.pdf.

[273] Mill Creek Watershed Council of Communities online, "Historical Timeline," accessed December 14, 2016. http://millcreekwatershed.org.

[274] Ibid.

[275] Ibid.

[276] "Waterproofing the Mill Creek Flood Plain," op. cit.

[277] US Army Corps of Engineers, Louisville District, "Mill Creek, Ohio Flood Damage Control Project," accessed June 11, 2016; http://www.lrl.usace.army.mil/Portals/64/docs/ReviewPlans/Review%20Plan%20(Approved)%20-%20Mill%20Creek,%20OH%202-20-14%20without %20names.pdf.

[278] Vermont Timber Works online, "Queen Post Truss," accessed June 19, 2016;
http://www.vermonttimberworks.com/our-work/timber-trusses/queen-post-truss/.

[279] Great Parks of Hamilton County online, "The Millstone Project," accessed December 14, 2016; http://blog.greatparks.org/2013/01/the-millstone-project/.

[280] The photo of the fire was taken by neighbor Robert Kettell.

[281] February 2016 interview with Mayor James Wolf.

[282] Cinci Golf online, "The Mill Course," accessed December 11, 2016; http://www.cincigolf.com/themillcourse/.

[283] Great Parks of Hamilton County online, "The Millstone Project," op. cit.

[284] Hill, *Descendants of Paul Hill and Rachel Stout...*, op. cit., p. 4. Wilson and Charles' grandmothers, Eliza Hendrickson Hill and Martha Hendrickson Hunt, were sisters.

[285] Kettell, "Flours and a Handsome Homestead," op. cit.

[286] Rhode and Hite, "Obed Hussey and his Ohio Test of the First Successful Reaper," op. cit. Eugene Brandt established his photography studio in Hamilton, Ohio around 1868.

[287] "Historic American Engineering Record No. OH-25," page 3. http://cdn.loc.gov/master/pnp/habshaer/oh/oh0400/oh0415/data/oh0415data.pdf.

[288] Email correspondence with Bob Mason, dated June 9, 2014.

[289] United States Federal Census, Year: *1850*; Census Place: *Springfield, Hamilton, Ohio*; Archive Collection Number: *T1159*; Roll: *12*; Line: *8*; Schedule Type: *Industry*.

[290] Logs that are de-limbed and prepared for sawing.

[291] Garber, *Waterwheels and Millstones...*, op. cit., p. 65.

[292] Theodore R. Hazen, *How the Roller Mills Changed the Milling Industry*, accessed December 5, 2016; http://www.angelfire.com/journal/millrestoration/roller.html.

[293] W. H. Alexander, *A Climatalogical History of Ohio, The Ohio State University Bulletin Vol. 28, No. 3, November 10, 1923*. (Columbus, OH: The Engineering Experiment Station of The Ohio State University. Printed by Ohio State Reformatory, Mansfield, 1924) pp. 308-309.

[294] Hamilton County Park drawings note that this section is an add-on.

[295] Garber, *Waterwheels and Millstones...*, op. cit., pp. 90-91.

[296] June 1981 interview with Herman Strassel, from Hamilton County Great Parks records.

[297] Hazen, *How the Roller Mills Changed the Milling Industry*, op. cit.

[298] Ibid.

[299] The History of Loudoun County, Virginia online, "Early 19th Century Milling and Wheat Farming," accessed June 11, 2016; http://www.loudounhistory.org/history/agriculture-mills-and-wheat.htm.

[300] June 1981 Herman Strassel interview.

[301] United States Federal Census, Year: *1900*; Census Place: *Springfield, Hamilton, Ohio*; Roll: *1283*; Page: *14A*; Enumeration District: *0318*; FHL microfilm: *1241283*.

[302] Registration State: *Ohio*; Registration County: *Hamilton*; Roll: *1832242*; Draft Board: *2*.

[303] National Association of Wheat Growers online, "Fast Facts," accessed December 16, 2016; *www.wheatworld.org/wheat-info/fast-facts/*.

[304] Ibid.

[305] May 1981 interview with Ralph Groff, from Hamilton County Parks records.

[306] *Cincinnati Enquirer*, Saturday, October 4, 1975, p. 20.

Section III

Appendixes

Timeline

1785
- December 25 – Samuel Stout Hill born

1793
- April 26 – Jediah Stout Hill born

1797
- January 31 – Eliza Hendrickson born

1806
- May 31 – Henry Rogers, Jr., born

c. 1814
- Samuel Hill and his family move to Hamilton County, Ohio. Samuel purchases land in Section 27 of Springfield Township.

1815
- April 29 – Jediah Stout Hill and Eliza Hendrickson are married.

1816
- January 27 – Rachel Maria Hill born

1817
- Samuel Hill and John Laboyteaux lay out the original town of Mount Pleasant.

1819
- Jediah Hill and family arrive in Springfield Township and settle in Section 28.

c. 1820–1822
- Jediah Hill and blacksmith John Lane design and build original sawmill.

1827
- March 11 – Samuel Hill dies

c. 1828
- Jediah Hill hires Henry Rogers, Jr., as hand in mill and on farm

c. 1830
- Jediah Hill and Henry Rogers, Jr., build large frame home overlooking mill site

1832
- September 22 – Henry Rogers, Jr., and Rachel Maria Hill are married.

1835
- Henry Rogers, Jr., on the design team that perfects the Obed Hussey Reaper

1838

- August 18 – The Hills and Rogers families commence their journey east.

c. 1839–1847

- Capital improvements to mill begun and completed

1840

- July 17 – Henry Rogers, Sr., dies

1843

- May 29 – Mary Jane Chadwick born
- December 29 – Wilson Thompson Rogers born

1844

- December 17 – Norbert Charles Hartmann born

1846

- January 15 – Nancy Gwaltney born

1850

- September 29 – Cati Roth Habig born
- Covered Bridge over Mill Creek completed

1854

- June 21 – Eliza Hendrickson Hill dies

1856

- April 30 – Amelia Engel born

1859

- July 4 – Jediah Hill dies

1860s

- Henry Rogers may have upgraded the mill and added a turbine.

1864

- May 10 – Wilson Rogers enlists for one hundred days' service in Union Army with his state militia unit, the 138th Ohio Volunteer Infantry.

1866

- March 15 – Wilson Rogers and Mary Jane "Mollie" Chadwick marry

1867

- July 26 – Harry Chadwick Rogers born
- August 3 – Norbert Charles Hartmann arrives in the United States.
- October 27 – Claude Clifton Groff born

1869

- September 27 – Walter Henry Rogers born
- October 22 – Charles Hunt born
- November 18 – Charles Hartmann and Cati Roth Habig marry

1870

- August 17 – Caroline Schwartz born

1872

- July ? – John Charles "Charlie" Hartmann born

1875

- April 15 – Henry and Rachel Rogers deed 194 acres and the mill property to Wilson Rogers.

1876

- November 11 – Henry and Rachel Rogers and Wilson and Mary Jane Rogers sell two acres of land to Francis Freiter.
- Wilson Rogers and family relocate to Denver, Colorado, seeking a better climate for relief of Mary Jane's lingering tuberculosis.

Timeline

1878

- July 12 – Katherine Rose Hartmann born

c. 1880

- Wilson Rogers and family return home to Springfield Township.

1881

- January 22 – Mary Jane Chadwick Rogers dies

1882

- November 9 – Wilson Rogers marries Nancy Gwaltney Smith.

1883

- August 10 – Charles Hartmann purchases mill at public auction

1884

- June 19 – Pearl Blaine Rogers born

August 29 – Maria Kunigunde Hartmann born

1886

- August 10 – Orpha Maria Rogers born

1888

- January 16 – Walter Rogers dies
- April 25 – Rachel Maria Hill Rogers dies

1889

- March 16 – Jay Ferris Rogers born
- November 25 – James Groff born

c. 1890

- Cati Hartmann dies

1891

- November 25 – Charles Hartmann and Amelia Engel Meyer marry.

1892

- September 28 – Josephine Hartmann born

1893

- December 17 – Ralph Groff born

1896

- May 23 – Clara Hartmann born
- December 1 – Henry Rogers, Jr., dies

1898

- Charles Hartmann upgrades business to steam-powered roller mill, retains lumber operation for the present
- September ? – Hildreth Hartmann born
- Name "Pride of the Valley" selected for Hartmann's premium grade flour

1899

- August 24 – Alvera Anna Ruther born

1900–1910

- Hartmann phases out lumber portion of business due to lack of raw materials and focus to merchant milling

1911

- Mill sold to C. C. Groff

1917

- March 25 Nancy Gwaltney Rogers dies

1918

- December 3 – Charles Hartmann dies in influenza epidemic

1921

- February 16 – James Groff dies

1922

- House built by Charles Hartmann next to mill destroyed by fire

1923

- April 24 – Ralph Louis Groff, Jr., born
- May 2 – Ralph Louis Groff, Jr., dies

1925

- Dorothy Elaine Groff born

1927

- December 3 – Caroline Schwartz Groff dies
- May 13 – Wilson Thompson Rogers dies
- Miriam Groff born

1929

- Joan Groff born

1934

- Ralph Groff takes over operation of mill

1936

- April 28 – Claude C. Groff dies

1941

- Charles Hunt pens his poem "The Old Covered Bridge"

1942

- Patricia Groff born

1943

- July 14 – Charles Hunt dies

1946

- January 23 – Amelia Hartmann dies

1952

- Mill closed as part of Army Corps of Engineers flood control project

1979–1980

- Interest in preserving the historic structure revived

1981

- October 25 – Mill destroyed by suspicious fire

1984

- January 17 – Ralph Groff dies

1993

- September 23 – Alvera Groff dies

II

Site Plan, Floor Plans, Elevations, and Cross Sections

The Mount Healthy Mill Recording Project was completed under conditions of Executive Order 11593 for the U. S. Army, Corps of Engineers, Louisville District, by the University of Cincinnati Department of Architecture. J. William Rudd acted as Project Supervisor, Bruce E. Goetzman was Field Director, and Eric M. Weckel was Student Architect.

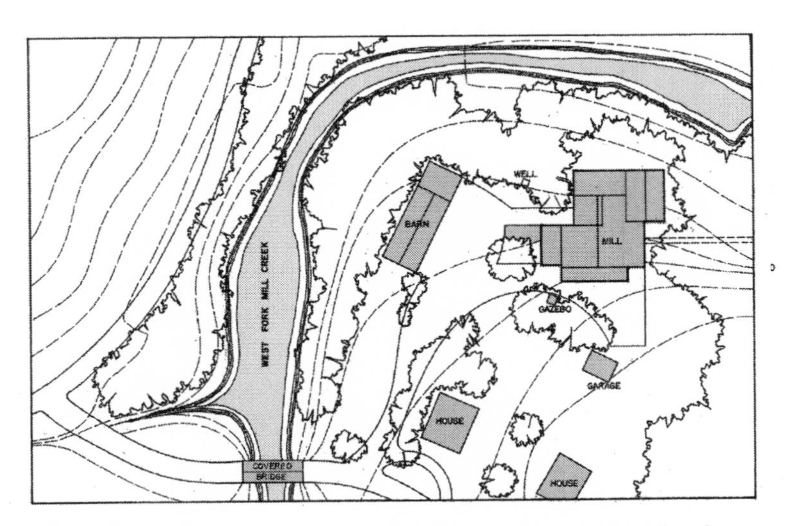

Figure A2.1A. Siteplan: excerpt of map of the Mount Healthy Mill and environs, including the barn, house, covered bridge, roads, and West Fork of Mill Creek.

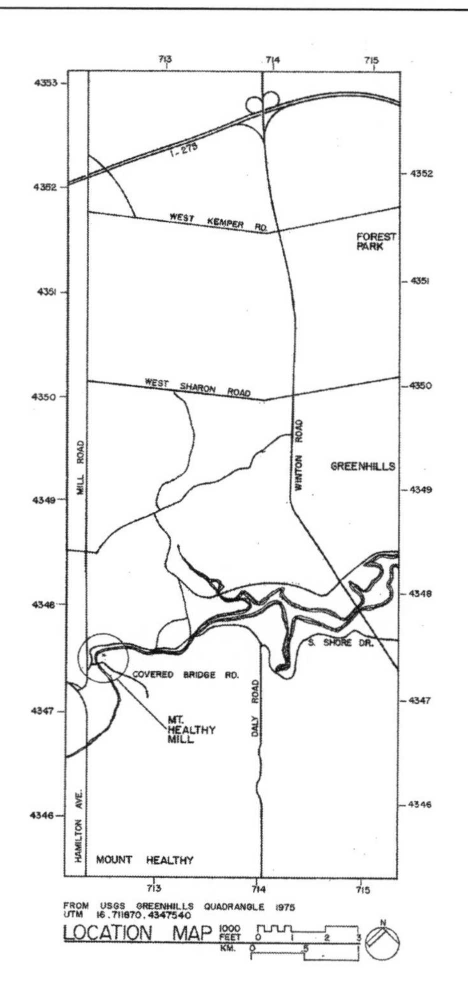

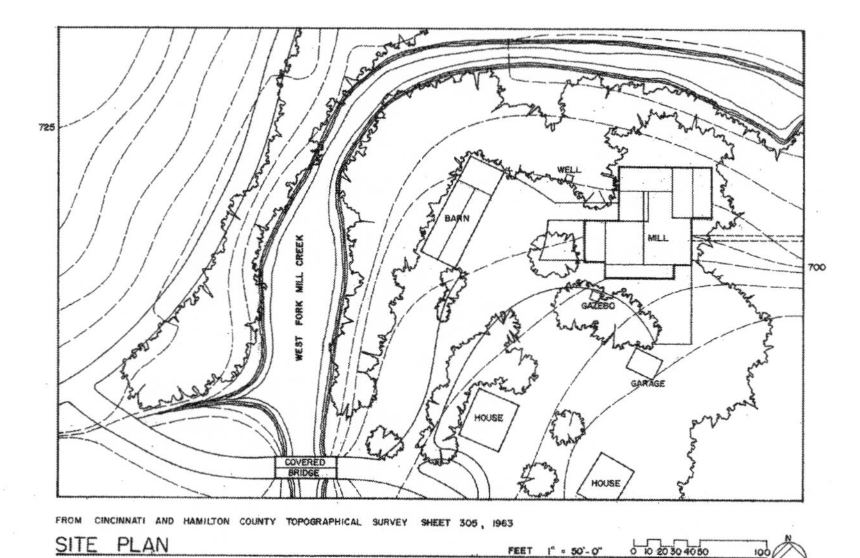

MT. HEALTHY MILL

THE MT. HEALTHY, OHIO, MILL WAS CONSTRUCTED c.1825 BY JEDIAH HILL AS A SAW MILL. IT IS THE ONLY MILL BUILDING OF ITS TYPE EXTANT IN THE CINCINNATI-HAMILTON COUNTY AREA. DURING THE MID-19th CENTURY THE MILL WAS OWNED BY HENRY ROGERS AND HIS SON, WILSON, PRIOR TO ACQUISITION BY CHARLES HARTMAN, c.1887. WITH DECREASING TIMBER RESOURCES, HARTMAN CONVERTED TO A GRIST MILL, c.1898. IN ADDITION TO MILLING COARSE FLOUR FOR NEARBY FARMERS, HE ALSO MILLED "PRIDE OF THE VALLEY" FLOUR FOR COMMERCIAL SALE IN CINCINNATI. THE PROPERTY WAS BOUGHT BY RALPH GROFF IN 1911 AND CONTINUED IN USE BY THE GROFF FAMILY UNTIL PURCHASED BY THE U.S. ARMY CORPS OF ENGINEERS AS PART OF THE WEST FORK LAKE FLOOD CONTROL PROJECT c.1952.

THE ORIGINAL SECTION OF THE BUILDING HAS AT LEAST THREE ADDITIONS, DATES UNDETERMINED. THE SUPER-STRUCTURE OF THE ORIGINAL BUILDING IS HAND-SHAPED WALNUT WITH A VARIETY OF JOINING TECHNIQUES. THE BUILDING IS PRESENTLY UNOCCUPIED AND SPECIFIC ROOM DESIGNATIONS ARE UNABLE TO BE DETERMINED.

THE MT. HEALTHY MILL RECORDING PROJECT WAS COMPLETED UNDER CONDITIONS OF EXECUTIVE ORDER — 11593 FOR THE U.S. ARMY, CORPS OF ENGINEERS, LOUISVILLE DISTRICT BY THE UNIVERSITY OF CINCINNATI, DEPARTMENT OF ARCHITECTURE; J. WILLIAM RUDD, PROJECT SUPSUPERVISOR; BRUCE E. GOETZMAN, FIELD DIRECTOR AND ERIC M. WECKEL, STUDENT ARCHITECT.

Figure A2.1B. Siteplan.

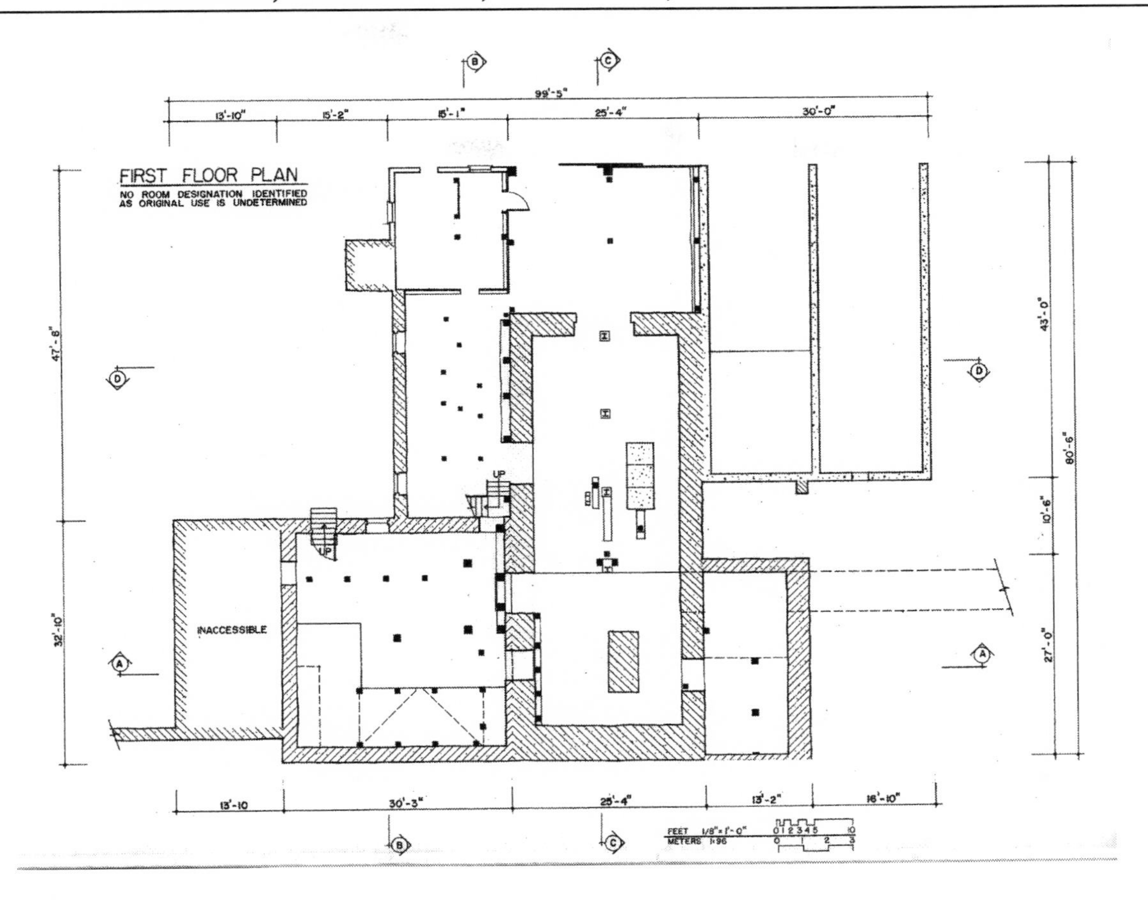

Figure A2.2. First floor.

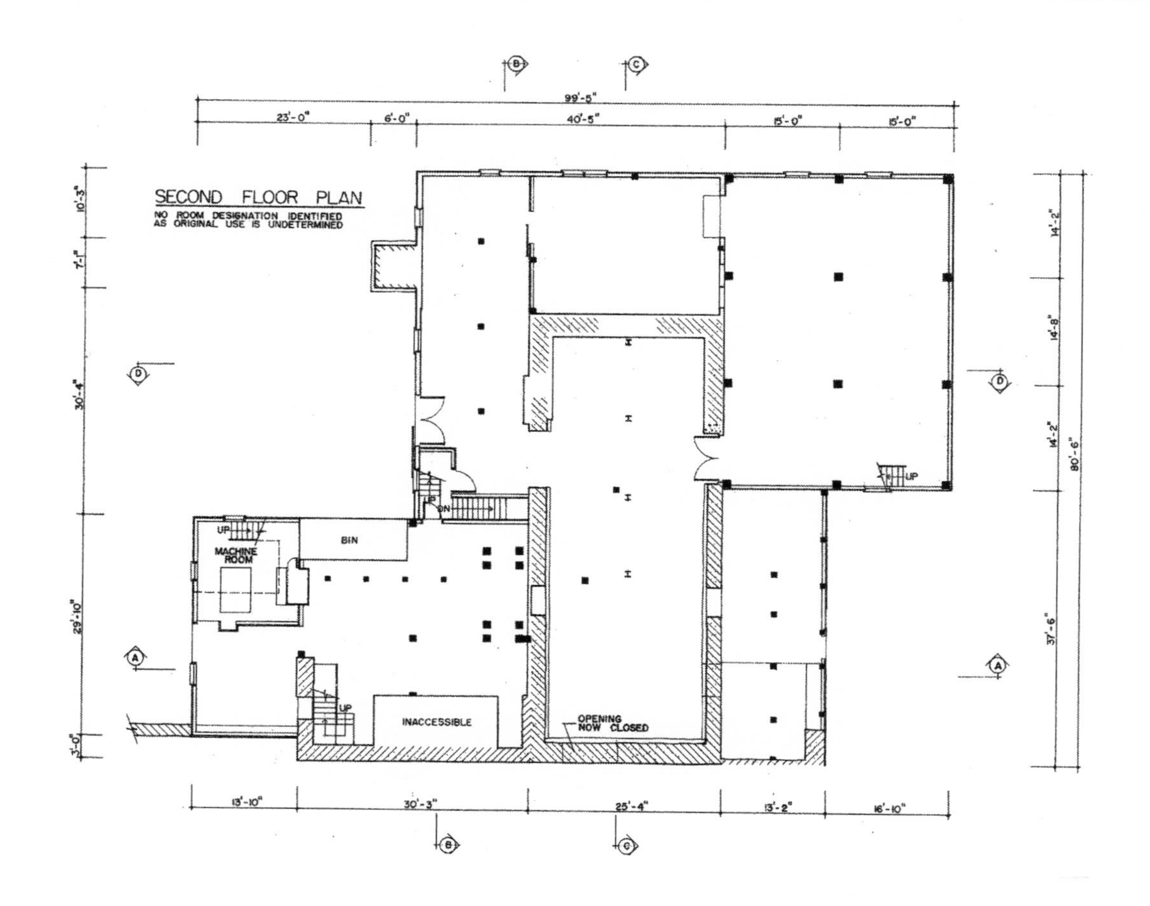

Figure A2.3. Second floor.

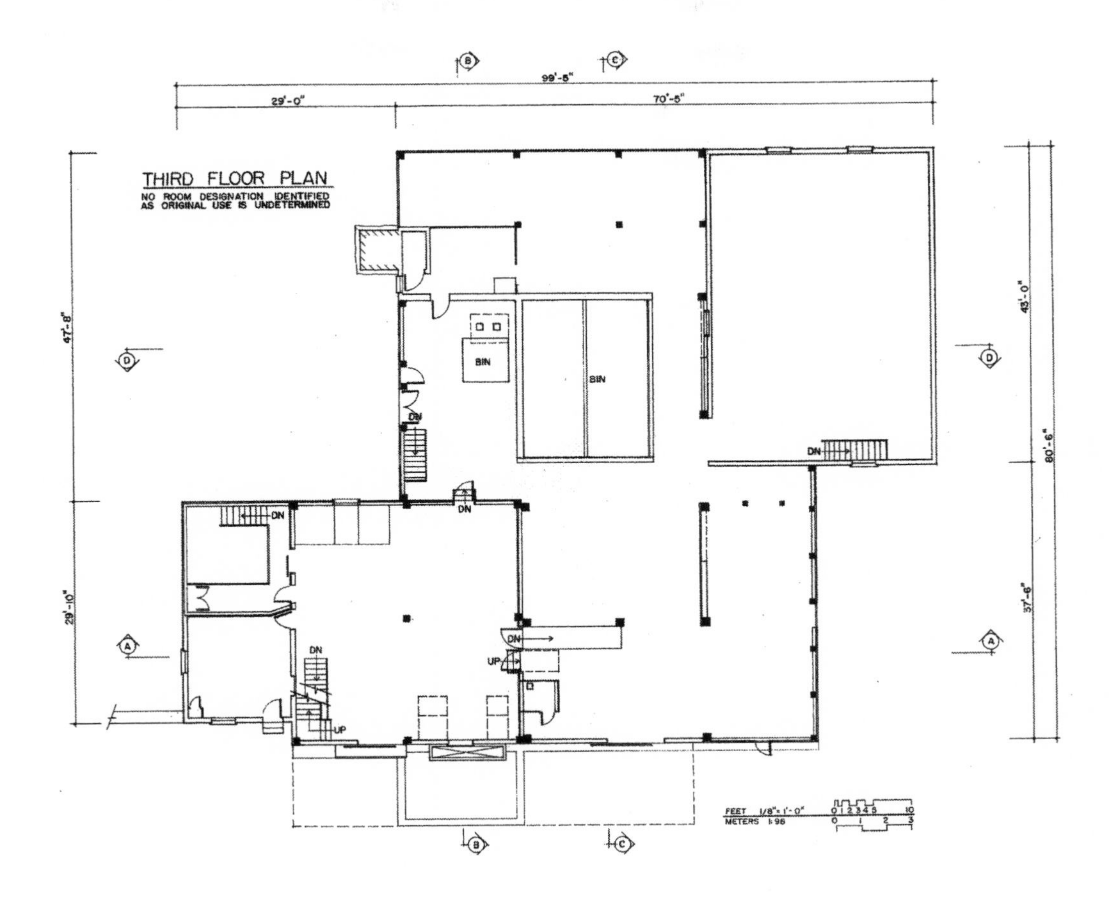

Figure A2.4. Third floor.

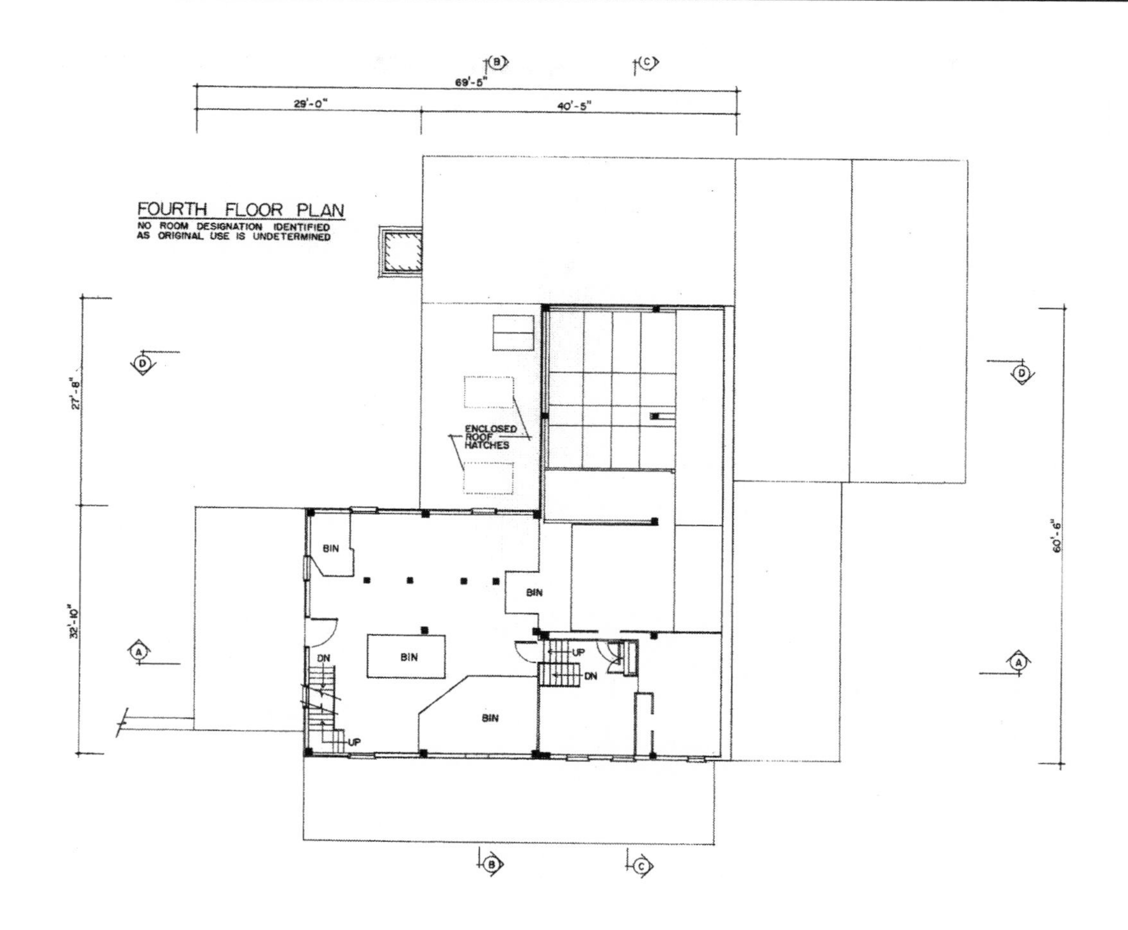

Figure A2.5. Fourth floor.

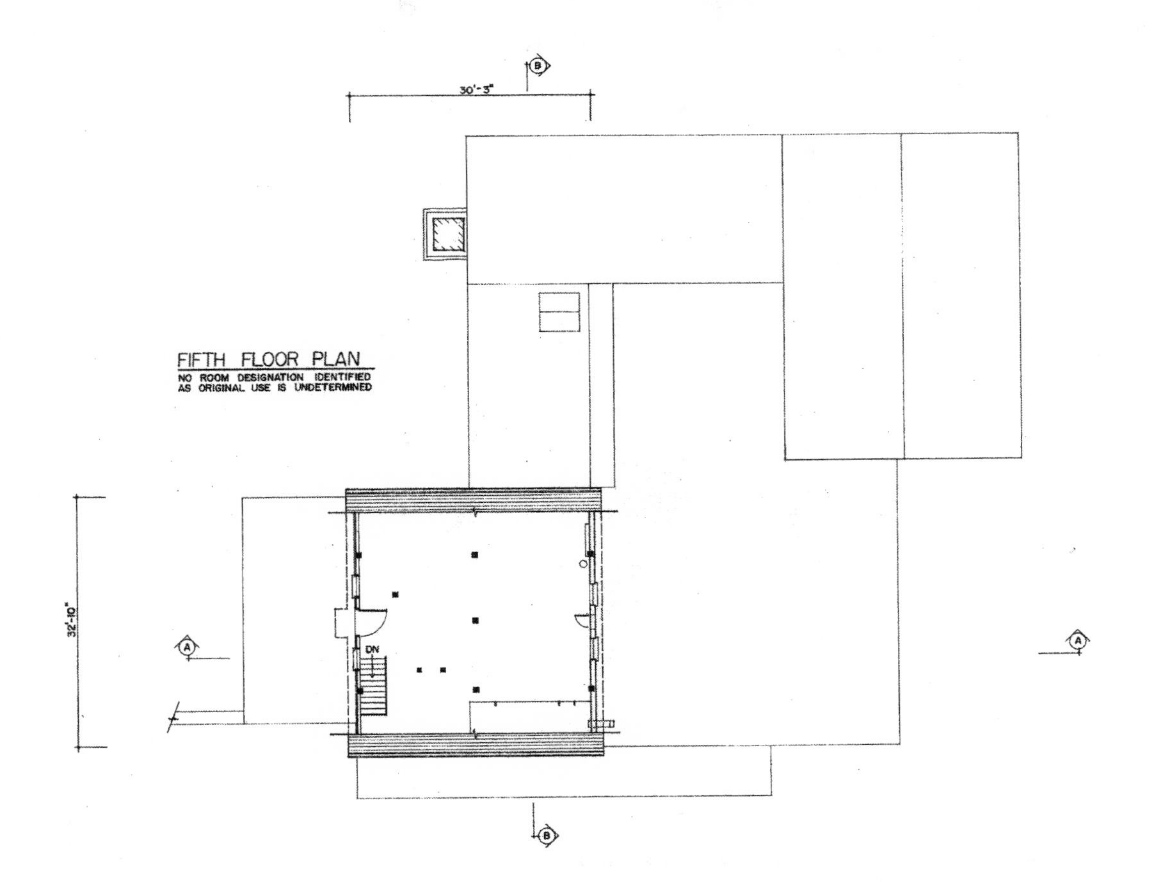

Figure A2.6. Fifth floor.

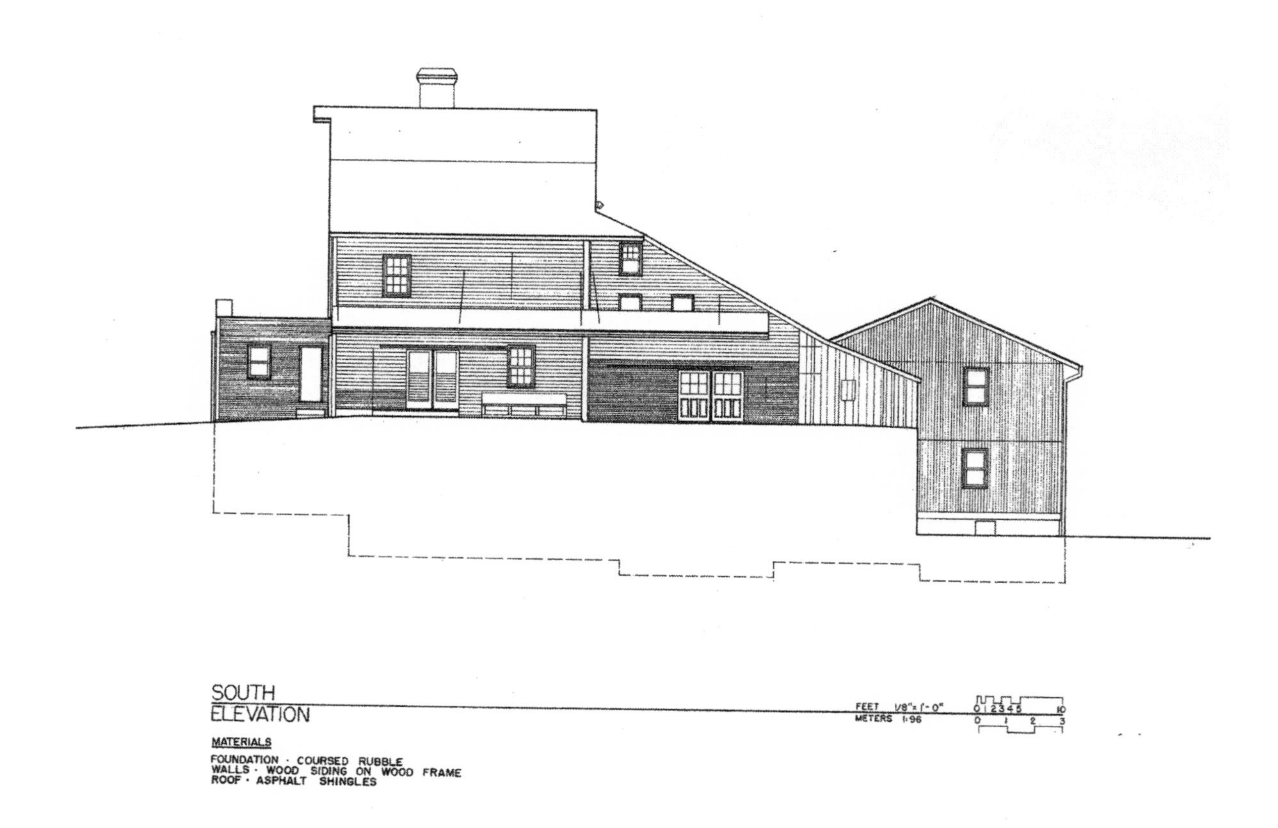

Figure A2.7. South elevation.

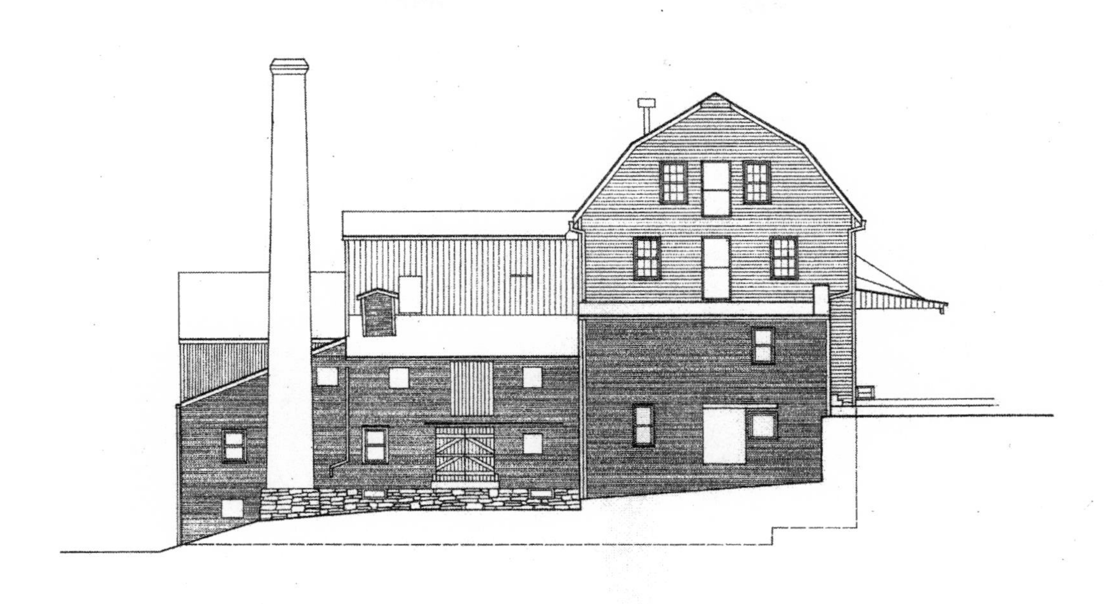

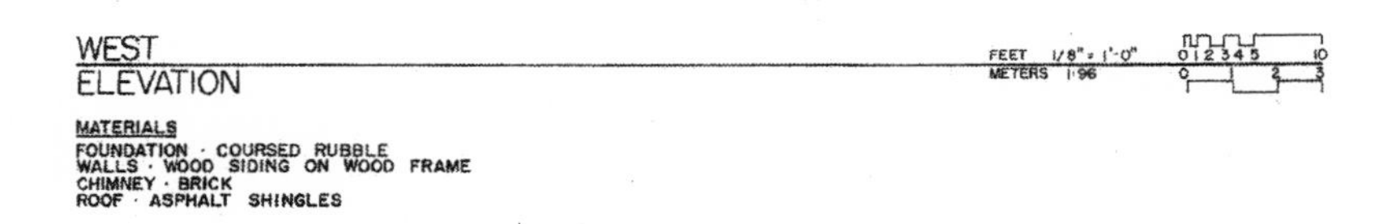

Figure A2.8. West elevation.

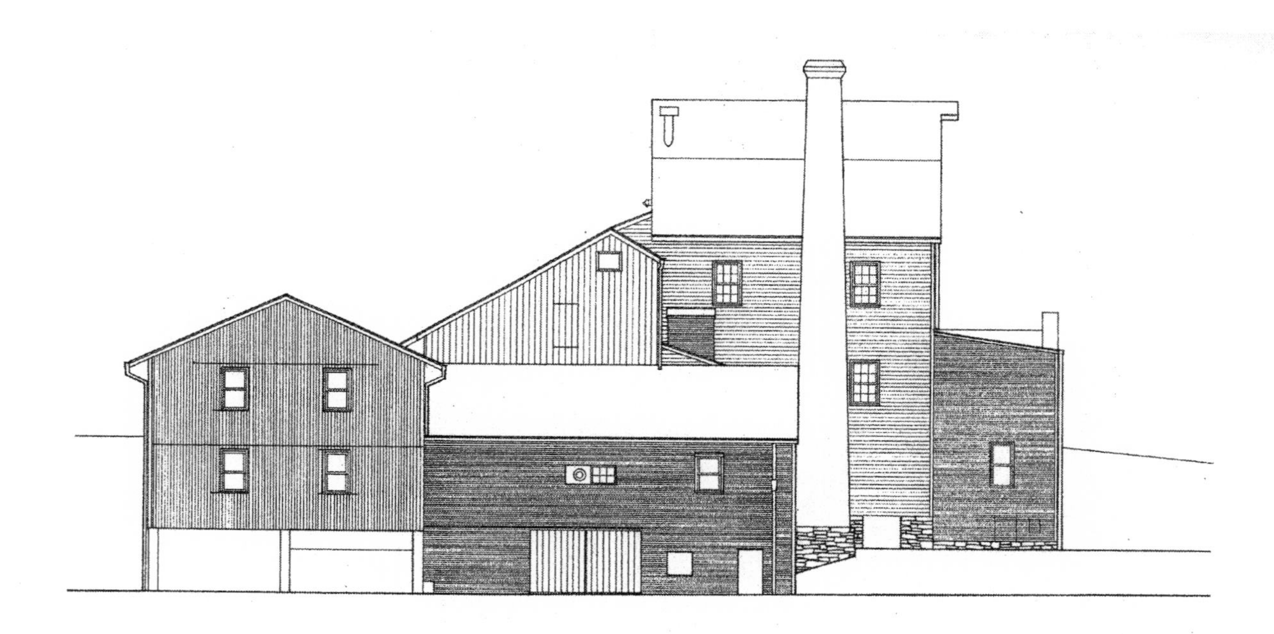

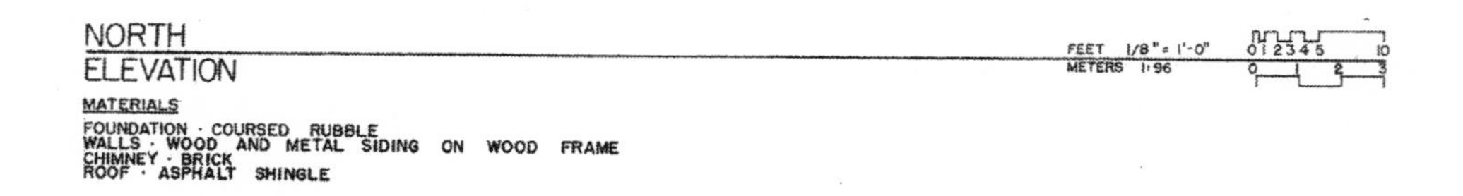

Figure A2.9. North elevation.

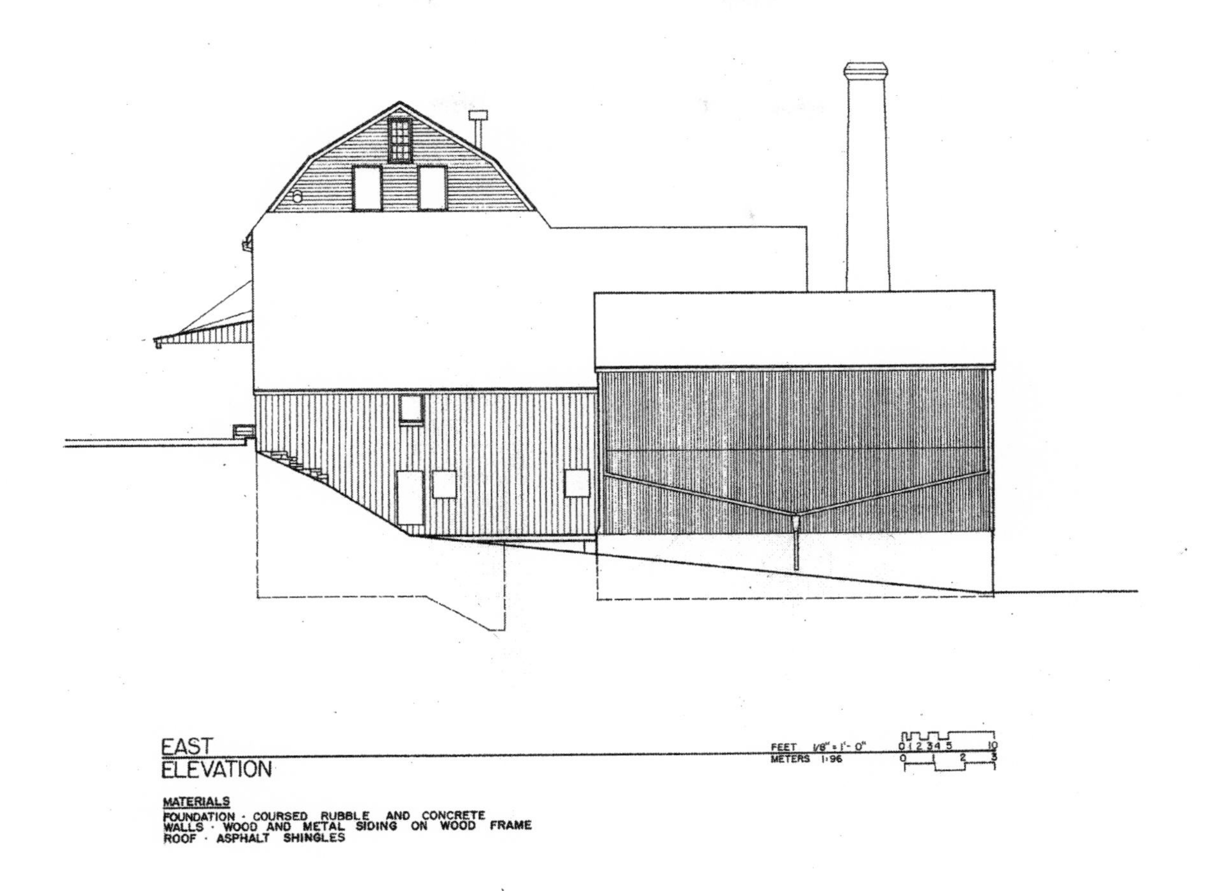

Figure A2.10. East elevation.

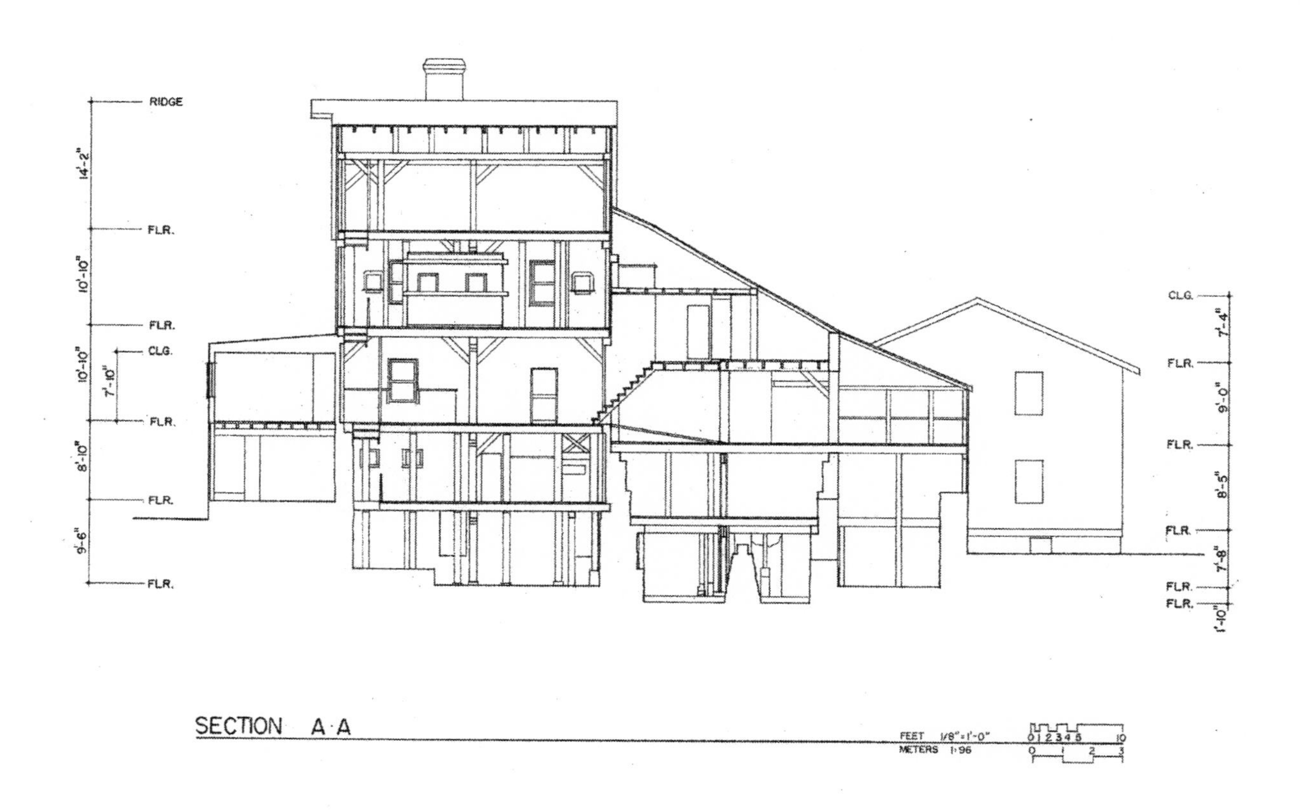

Figure A2.11. Cross section A-A.

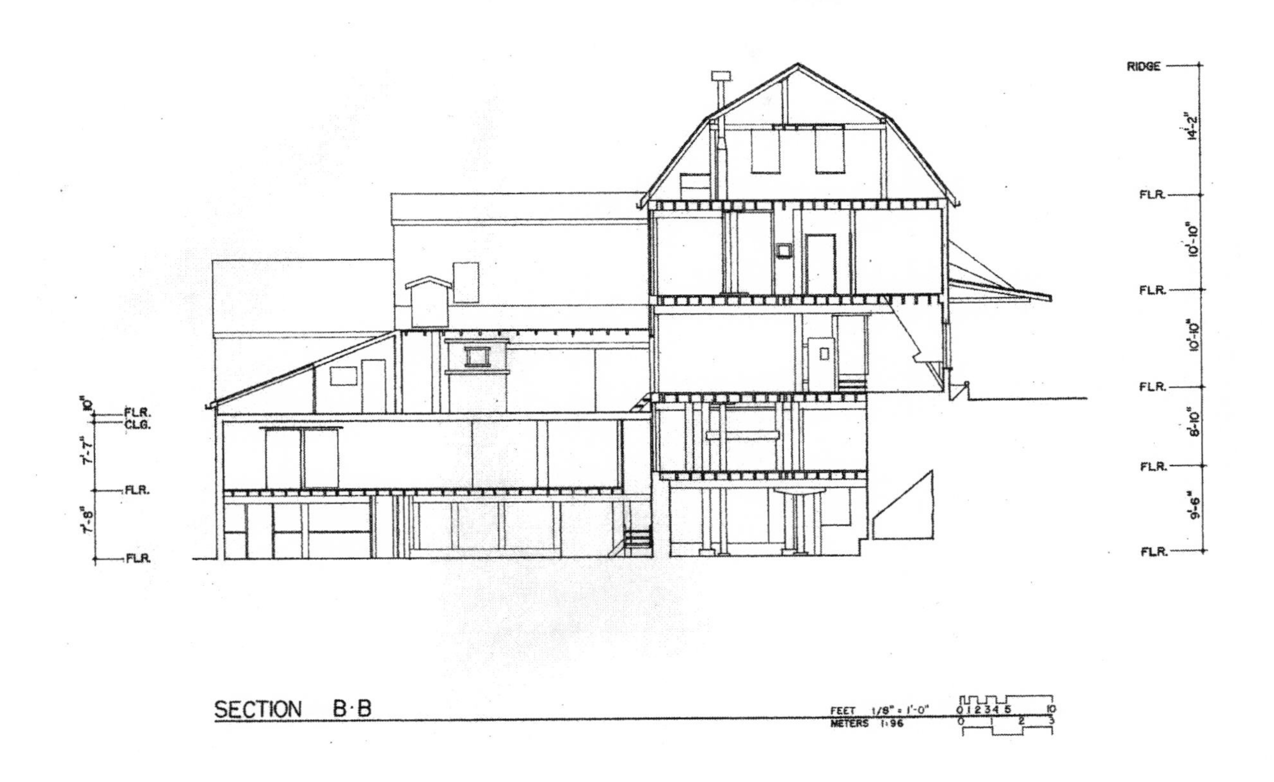

Figure A2.12. Cross section B-B.

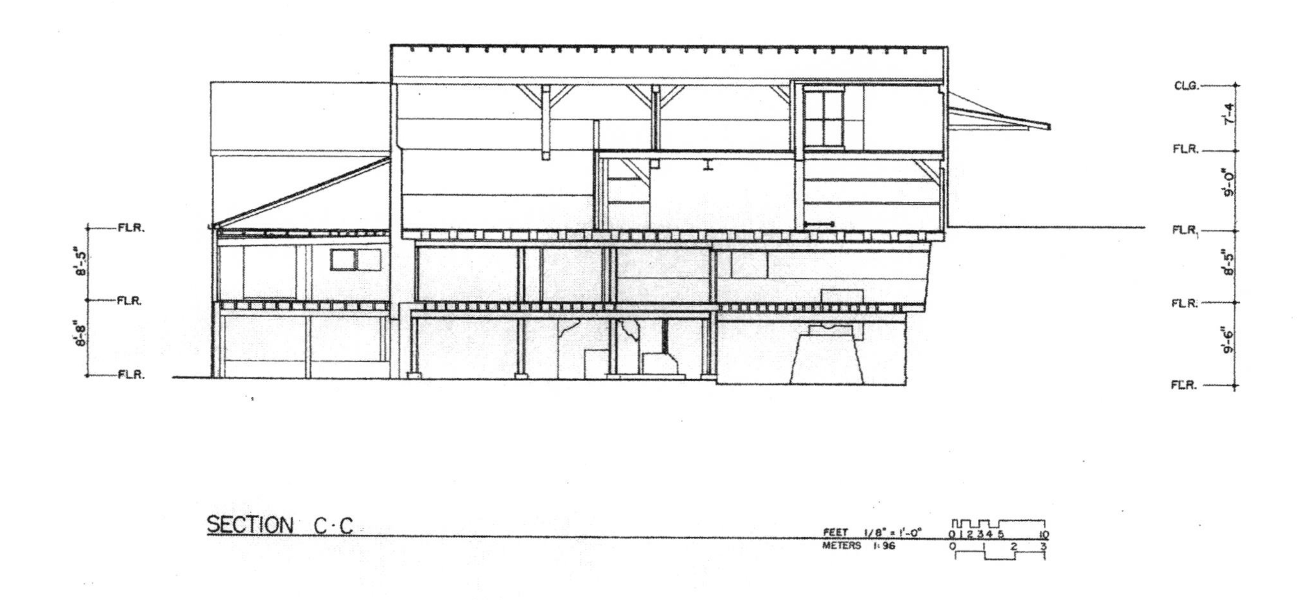

Figure A2.13. Cross section C-C.

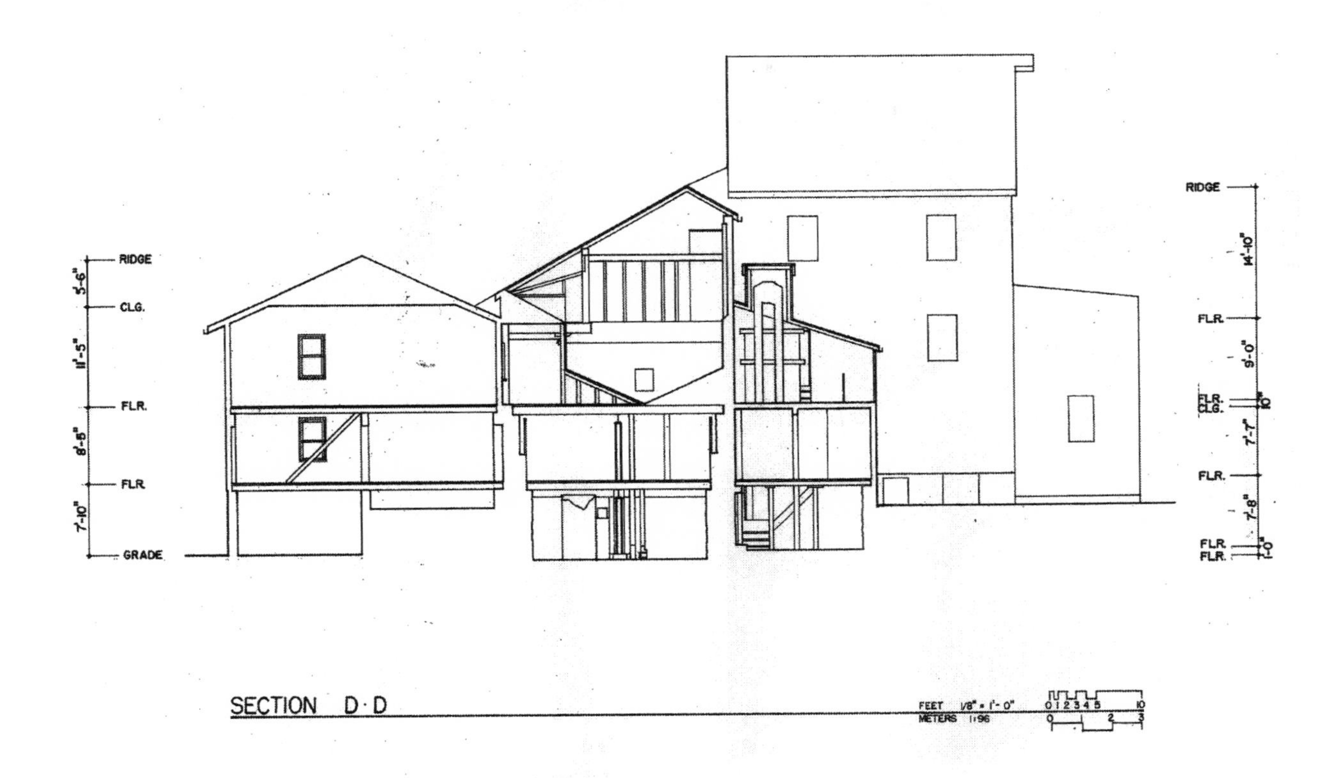

Figure A2.14. Cross section D-D.

III

Historic American Engineering Record Photographs

These photographs of the exterior and interior of the mill were taken as part of the written, historical, and descriptive data compiled about the Mount Healthy Flour Mill by the U. S. Department of the Interior.

File HAER No. OH-25, containing the written account of the mill and its history, was completed and transmitted in 1986, five years after the mill was destroyed by fire. That report is supplemented by File HAER No. OH-31, which contains the photographs taken of the interior and exterior of the mill.

Figure A3.1. South elevation.

Figure A3.2. West elevation.

Figure A3.3. West elevation of rear wing.

Figure A3.4. North and west elevations and well house.

Figure A3.5. North elevation.

Figure A3.6. East elevation.

Figure A3.7. Detail of canopy on south elevation.

Figure A3.8. Mill race exit.

Figure A3.9. Detail of post and stone wall, lower level, center room.

Figure A3.10. Stone pier that supported original water wheel, lower level, south room.

Figure A3.11. Mill race exit, lower level, south room.

Figure A3.12. Second (main) floor, middle room.

Figure A3.13. Second (main) floor, machine room.

Figure A3.14. Third floor, bin room.

Figure A3.15. Detail of post and beam construction, third floor bin room.

Figure A3.16. Detail of post, beam, and bracing, third floor.

Figure A3.17. Detail of post, beam, and bracing, fourth floor.

Figure A3.18. Fifth floor.

Figure A3.19. Detail of grain bin.

Figure A3.20. Detail of post and bracing.

IV

Credits for Photographs and Other Images

Bredenfoerder, John H.: 3.1

Brigode, Winona Rogers: 11.3, 11.6, 11.10

Pat Brown Photography: 2.1, 19.2

Buck, Patricia Groff: 13.3, 13.4, 13.5, 13.8

Delaware Department of Transportation: 19.15

Glendale Heritage Preservation: 4.7

Great Parks of Hamilton County: 17.2, 18.1, 18.3, 18.4, 19.3

Gregory, Brent: 16.2, 16.4, 16.5, 17.5

Hagaman, Steve: 19.12, 19.13

Hardwick, Kevin and Cindy: 17.1, 17.3, 17.4

Jesionowski, Tom: 12.1, 12.5, 12.7

Lawson, Tracy: I.2, 1.2, 1.3, 3.2, 7.3, 8.3, 8.4, 9.4, 10.1, 11.1, 11.2, 11.4, 11.8, 11.9, 11.13, 13.7, 15.1, 15.2, 15.3, 16.1, 16.6, 17.5, 18.2, 19.5, 19.16, 19.17, 19.18, 19.20

Mount Healthy Historical Society: 1.4, 4.4, 7.6, 7.8, 8.1, 8.2, 10.2, 12.6, 13.1, 13.6

Mussey, Sandra Chadwick: 9.5

Neço, Andrea: 19.4, 19.6, 19.7, 19.14, 19.19

Prout, Don: 12.2, 12.3, 12.4, 14.2, 14.3, 14.4, 19.1

Rieck, Florence Rogers: 13.2

Rogers, Craig and Walter: 8.5, 16.7

Sharonville United Methodist Church: 7.7

Image on Creative Commons: 5.2

Images in the Public Domain: I.1, 1.1, 1.5, 3.3, 4.1, 4.2, 4.3, 4.5, 4.6, 5.1, 5.3, 5.4, 7.1, 7.2, 7.4, 7.5, 7.9, 8.6, 8.7, 8.8, 9.1, 9.2, 9.3, 19.8, 19.9, 19.10, 19.11

Images from author's private collection: 10.3, 11.5, 11.7, 11.11, 11.12, 11.14, 14.1, 16.3

Bibliography

A roster of Revolutionary Ancestors of the Indiana Daughters of the American Revolution: commemoration of the United States of America bicentennial, July 4, 1976. Evansville, IN: Unigraphic, 1976.

Alexander, W. H. *A Climatalogical History of Ohio, The Ohio State University Bulletin Vol. 28, No. 3, November 10, 1923*. Columbus, OH: The Engineering Experiment Station of The Ohio State University. Printed by Ohio State Reformatory, Mansfield, 1924.

Benson, B. C. "A Northern California Herd," *Holstein-Friesian World*. Volume 18, Issue 2, December 10, 1921.

Breyfogle, William A. *Wagon Wheels: A Story of the National Road*. New York, NY: Aladdin Books, 1956.

Cahill, Laura Schmidt and David L. Mowery, *Morgan's Raid Across Ohio: The Civil War Guidebook of the John Hunt Morgan Trail*. Columbus, OH: The Ohio Historical Society, 2014.

Cheney, Frank Woodbridge. *Subject: The Underground Railroad*. Cheney Family Manuscript Collection Ms 92876, Connecticut Historical Society, Hartford, Connecticut. 1901.

Cist, Charles. *Sketches and Statistics of Cincinnati in 1851*. Cincinnati, OH: Wm. H. Moore & Co., Publishers, 1851.

"Citizens Hope to Save Covered Bridge," *Cincinnati Enquirer*, Sunday February 12, 1956. Author's collection.

Commissioners to Examine Claims Growing Out of the Morgan Raid. *Report of the Commissioners of Morgan Raid claims: to the Governor of the state of Ohio, December 15th, 1864*. Columbus, OH: R. Nevins, State Printer, 1865.

Cooley, Eli F. and William S. Cooley. *Genealogy of Early Settlers in Trenton and Ewing "Old Hunterdon County," New Jersey*. Baltimore, MD: Genealogical Publishing Co, Inc., 1977.

Deats, Hiram E., ed. *Marriage Records of Hunterdon County, New Jersey, 1795-1875*. Flemington, NJ: J. E. Deats, Publisher, 1918.

Dexter, Jim. "County Wants Old Gristmill, Now Used As Dump, Preserved," *Cincinnati Enquirer*, September 14, 1981.

Evans, Oliver. *The Young Mill-Wright & Miller's Guide with Foreword by Eugene S. Ferguson, Reprinted from the First Edition, 1795*. Wallingford, PA: The Oliver Evans Press, 1990.

Foraker, Joseph B., Governor, and James S. Robinson, Secretary, and H. A. Axline, Adjudicant-General. *Official Roster of the Soldiers of the State of Ohio in the War of the Rebellion, 1861-1866, Vol. VIII 110th-140th Regiments, Infantry*. Cincinnati, OH: The Ohio Valley Press, 1888.

Ford, Henry A., A. M., and Mrs. Kate B. Ford. *History of Hamilton County, Ohio with Illustrations and Biographical Sketches.* Cleveland, OH: L. A. Williams & Col, Publishers, 1881.

Garber, D. W. *Waterwheels and Millstones: A History of Ohio Gristmills and Milling, Historic Ohio Buildings Series Volume 2.* Columbus, OH: The Ohio Historical Society, 1970.

Geier, Herbert J. New Burlington Civic Association Sesquicentennial Celebration Committee. *New Burlington — 100 Years Ago: New Burlington's Sesqui-Centennial May 1816-May 1966.*

Greeno, Follett L., ed., *Obed Hussey Who, of All Inventors, Made Bread Cheap.* Rochester, NY: Follett L. Greeno, 1912.

"Hall Road Has Long History," *The Forest Park News*, Tuesday, February 4, 1969. Glendale Heritage Preservation collection, Glendale, Ohio.

Halstead, Murat. *The Paddy's Run Papers.* Published by M. Halstead. Unknown Binding, 1895.

Harlow, Lindley. "Western Travel, 1800-1820." *The Mississippi Valley Historical Review,* Volume 6, Number 2. September 1919.

Harper, Glenn, and Doug Smith. *The Traveler's Guide to the Historic National Road in Ohio.* Columbus, OH: The Ohio Historical Society, 2005.

Harper's Weekly Magazine, Volume VII, Number 343, July 25, 1863.

Hartmann, Charles. Unpublished journal. Author's collection.

Hebeler, D. J. *Mt. Healthy of Former Days: A History Compiled from Written and Oral Testimony*. Unpublished manuscript, The Public Library of Cincinnati and Hamilton County collection.

Hedeen, Stanley. "Waterproofing the Mill Creek Flood Plain." *Queen City Heritage*, Spring 1998.

Hill, Leonard, and Louise Hill. *Descendants of Paul Hill and Rachel Stout through Charles Hill and of Moses Edwards and Desire Meeker through Uzal Edwards.* Unpublished manuscript, 1953. Author's private collection.

Holt, Michael F. *The Rise and Fall of the American Whig Party: Jacksonian Politics and the Onset of the Civil War.* New York, NY: Oxford University Press, 1999.

Homestead, Melissa J. *American Women Authors and Literary Property, 1822-1869.* New York, NY: Cambridge University Press 2005.

Horwitz, Lester V. *The Longest Raid of the Civil War*. Cincinnati, OH: Farmcourt Publishing, 2001.

Hover, John C., ed. *Memoirs of the Miami Valley in Three Volumes, Illustrated*. Chicago, IL: Robert O. Law Company, 1919.

Howe, Henry, LL.D. *Historical Collections of Ohio in Two Volumes: An Encyclopedia of the State.* Cincinnati, OH: C. J. Krehbiel & Co., Printers and Binders, 1907.

Bibliography

Jones, Grace Estel. *The Story of New Burlington 1816-1922*. Hilltop Publishing Co., Printers. Mt. Healthy, Ohio, 1922. Author's private collection.

Kettell, Carolyn. "Flours and a Handsome Homestead." *Cincinnati Enquirer Tristate Magazine*, Sunday, October 2, 1988.

Larkin, David, *Mill: The History and Future of Naturally Powered Buildings*. New York, NY: Universe Publishing, 2000.

Lawson, Tracy. *Fips, Bots, Doggeries, and More: Explorations of Henry Rogers' 1838 Journal of Travel from Southwestern Ohio to New York City*. Granville, OH: McDonald & Woodward Publishing Co., 2012.

Longworth, Thomas. *Longworth's American Almanac, New-York Register and City Directory for 1839*. New York, NY: 118 Nassau Street, 1839.

Loomis, Greg. "County park district trying to save historic mill," *Northern Hills Press*, July 1, 1981.

Map of Hamilton County, Ohio, by Wm. D. Emerson, Cincinnati, OH: C. S. Williams and Son, 1847.

Map of Hamilton County, Ohio : exhibiting the various divisions and sub divisions of land with the name of the owners & number of acres in each tract together with the roads, canals, streams, towns &c. throughout the county. Compiled from actual surveys by A. W. Gilbert. Cincinnati, OH: Engraved by Ed. O. Reed, 1856.

Map of Hunterdon County, New Jersey, Entirely from Original Surveys by Samuel J. Cornell, Surveyors. Camden, NJ: Lloyd Van Der Veer and J. C. Cornell, Publishers, 1851.

Map of Mercer County, New Jersey, Entirely from Original Surveys by J. W. Otley and J. Keily, Surveyors. Camden, NJ: Lloyd Van Der Veer, Publisher, 1849.

Mayer, Mabel Watkins. "Into the Breach: Civil War Letters of Wallace W. Chadwick." *The Ohio State Archaeological and Historical Quarterly* Volume 52, Number 2, April-June, 1943.

McClure, Stanley. *The descendants of William Sater (1793-1849), Crosby township, Hamilton County, Ohio*. Cincinnati, OH: S.W. McClure, Publisher, 1986.

McCormick, Richard P. "The 'Ordinance' of 1785," *William and Mary Quarterly*, Volume 50, Issue 1, January 1993.

McCutcheon, Mark. *The Writer's Guide to Everyday Life in the 1800s*. Cincinnati, OH: Writer's Digest Books, 1993.

Niesen, Marjorie N. *Images of America: St. Bernard*. Charleston, SC: Arcadia Publishing, 2011.

Official roster of the soldiers of the State of Ohio in the War of the Rebellion, 1861-1866 / compiled under the direction of the Roster Commission. *Akron, OH :Werner Co., 1886-1895*.

One Square Mile 1817-1992. Mt. Healthy, OH: Mt. Healthy Historical Society, 1992.

Parrish, Charles E., Historian for U.S. Army Corps of Engineers. *United States Department of the Interior, National Park Service National Register of Historic Places Inventory — Nomination Form: Mt. Healthy Mill*. July, 1980. Author's private collection.

Peterson, Arthur J., Chairman. Mount Healthy Sesquicentennial Celebration Committee. *Once Upon a Hilltop: Mount Healthy Area Sesquicentennial 1817-1967.*

Reminiscental, unpublished manuscript of autobiography of Clark Lane, on file at Butler County Historical Society, Hamilton, Ohio.

Rhode, Robert T., and Leland Hite. "Obed Hussey And His Ohio Test of the First Successful Reaper," *Engineers & Engines Magazine*, Volume 60, Number 4, October-November 2014.

Rothbard, Murray. *A History of Money and Banking in the United States: The Colonial Era to World War II.* Edited with an Introduction by Joseph T. Salerno. Auburn, AL: The Ludwig Von Mises Institute, 2002.

Schaefer, Walt. "Park Officials Ponder Life Without Grist Mill," *Cincinnati Enquirer,* Tuesday, October 27, 1981.

Scheurer, L. Miami Purchase Association. *Ohio Historic Inventory #HAM-1347-49: C. C. Groff Flour Mill.* Author's private collection.

Sloane, Eric. "The Mills of Early America," *American Heritage Magazine*, Volume 6, Number 6, October 1955.

Smiddy, Betty Ann, ed. *A Little Piece of Paradise...College Hill, Ohio.* 2nd edition. Cincinnati, OH, 2008.

Smith, Elmer L. *Grist Mills of Early America and Today.* Lebanon, PA: Applied Arts Publishers, 1978.

Snodgrass, Mary Ellen. *The Underground Railroad: An Encyclopedia of People, Places, and Operations.* London: Taylor & Francis, 2008.

Stewart, Dr. David, and Dr. Ray Knox. *The Earthquake America Forgot: 2000 Temblors in Five Months...And It Will Happen Again!* Marble Hill, MO: Gutenberg-Richter Publications, 1995.

Sussman, Lawrence, "Old covered bridge to be renovated," *Cincinnati Enquirer,* date unknown, 1977. Author's collection.

Taylor, William A., Secretary of the Ohio Society S. A. R., "Living Ohio Sons of Revolutionary Sires," *The Ohio Magazine*, Volume 2, January, 1906.

Titus' Atlas of Hamilton County, Ohio Entirely of Original Surveys by R. R. Harrison, C. E. Cincinnati, OH: C. O. Titus, Publishers, 1869.

Twentieth Annual Report of the Department of Inspection of Workshops, Factories, and Public Buildings to the Governor of the State of Ohio for the year 1903. Springfield, OH: The Springfield Publishing Company State Printers, 1904.

U. S. Department of the Census, *Marriage and divorce 1867-1906.* Ann Arbor, MI: University of Michigan Library, 1908.

Vickery, Amanda. "Golden Age to Golden Spheres? A Review of the Categories and Chronology of English Women's History," *The Historical Journal,* 36.1, June 1993.

Weisenberger, Francis P. *The History of the State of Ohio, Vol. III The Passing of the Frontier, 1825-1850.* Carl Wittke, ed. Columbus, OH: Ohio State Archaeological and Historical Society, 1941.

White, C. Albert. *A History of the Rectangular Survey System*. U. S. Department of the Interior, Bureau of Land Management, 1983.

White, Ronald C. Jr. *A. Lincoln: A Biography*. New York, NY: Random House, 2009.

Wilson, Sue Korn, and Kathleen Mulloy Tamarkin. *Images of America: Mt. Healthy*. Charleston, SC: Arcadia Publishing, 2008.

Wittke, ed., and Eugene H. Rosebloom. *The History of the State of Ohio: The Civil War Era 1850-1873*. Columbus, OH: Ohio State Archaeological Society 1944.

Videos, Websites, and Web Pages

About.com, "The Agricultural Revolution," accessed April 27, 2016; http://inventors.about.com/od/indrevolution/a/AgriculturalRev.htm.

About.com, "Cyrus McCormick, Inventor of the Mechanical Reaper," accessed December 13, 2016; http://inventors.about.com/od/famousinventions/fl/Cyrus-McCormick-Inventor-of-the-Mechanical-Reaper.htm.

Agriculture in the Classroom online, Growing a Nation: The Story of American Agriculture "Historical Timeline — Farm Machinery & Technology" accessed April 2, 2016; https://www.agclassroom.org/gan/timeline/farm_tech.htm.

Ancestry.com, *Selected U.S. Federal Census Non-Population Schedules, 1850-1880* [database on-line]. Provo, UT, USA: Ancestry.com Operations, Inc., 2010.

Ancestry.com, *Revolutionary War Pension and Bounty-Land Warrant Application Files, 1800-1900* [data-base online]. Provo, UT, USA. Ancestry.com Operations, Inc. 2010.

Arnett, Karen. Hamilton Avenue Road to Freedom online, accessed December 16, 2016; http://hamiltonavenueroadtofreedom.org/.

Bellis, Mary. About online, "Machines to Cut Grains" accessed April 1, 2016; http://inventors.about.com/od/rstartinventions/a/reaper.htm.

Bellis, Mary. About online, "Cyrus McCormick Inventor of the Mechanical Reaper," accessed May 3, 2016; http://inventors.about.com/od/famousinventions/fl/Cyrus-McCormick-Inventor-of-the-Mechanical-Reaper.htm.

Built in America online, Historic American Engineering Record HAER OH-25 Mount Healthy Mill, accessed April 12, 2016; http://cdn.loc.gov/master/pnp/habshaer/oh/oh0400/oh0415/data/oh0415data.pdf.

Corathers, Robin. Cincinnati.com, "Life Blooming in Mill Creek," accessed December 13, 2016; http://www.cincinnati.com/story/opinion/contributors/2015/03/03/life-blooming-mill-creek/24325975/.

Castiglia, Carolyn. Babble online, "Thoughts on Love, Dating, Marriage, and Divorce from 1870," accessed June 5, 2016; http://www.babble.com/mom/thoughts-on-love-dating-marriage-and-divorce-from-1870/.

Cincinnati Fire Apparatus Resource online, "North College Hill Fire Department," accessed June 11, 2106; http://www.cincyfireapparatus.com/northcollegehill.html.

Cincinnati Friends Meeting online, "Historical Archive: Newspaper Clippings," accessed December 6, 2016; http://cincinnatifriends.org/about/meetinghistory.html.

Cincinnati Museum online, "Waterproofing the Mill Creek Flood Plain," http://library.cincymuseum.org/topics/f/files/1937flood/wat-015.pdf.

Cinci Golf online, "The Mill Course," accessed December 11, 2016; http://www.cincigolf.com/themillcourse/.

Digital History online, "Percentage of American Labor Force in Agriculture," accessed April 2, 2016; http://www.digitalhistory.uh.edu/disp_textbook.cfm?smtID=11&psid=3837.

Explore Pennsylvania History online, "Overview: Agriculture and Rural Life," accessed April 27, 2016; http://explorepahistory.com/story.php?storyId=1-9-4.

Family Search online, "Ohio Hamilton County Records, 1791-1994, Land and Property Records, Deed and Mortgage Index 1794-1859," accessed December 16, 2016; https://familysearch.org/ark:/61903/3:1:3QS7-L92Z-C962-D?mode=g&i=357&cc=2141016.

Gates, Henry Louis Jr., The Root online, "Who Really Ran the Underground Railroad?" accessed February 14, 2016; http://www.theroot.com/articles/history/2013/03/who_really_ran_the_underground_railroad.html.

Gazit, Chana., dir. *The Forgotten Plague: Tuberculosis in America*. WGBH Educational Foundation, 2015. Accessed April 12, 2016; http://www.pbs.org/wgbh/americanexperience/films/plague/.

George Washington's Mount Vernon online, "Ten Facts About the Gristmill," accessed December 6, 2016; http://www.mountvernon.org/the-estate-gardens/gristmill/ten-facts-about-the-gristmill/.

The German Americans online, "Why Germans Left Home," accessed April 19, 2016; http://maxkade.iupui.edu/adams/chap2.html.

Gold, David M. *Eminent Domain and Economic Development: The Mill Acts and the Origins of Laissez-Faire Constitutionalism*, accessed December 13, 2016; http://direct.mises.org/sites/default/files/21_2_5.pdf.

Great Parks of Hamilton County online, "The Millstone Project." Accessed December 10, 2016; http://blog.greatparks.org/2013/01/the-millstone-project/.

Gunn, Michael. Share online, "18 Revolutionary War Patriots finally honored at Wesleyan Cemetery." Accessed April 12, 2016; http://local.cincinnati.com/share/story/225417.

Hazen, Theodore R. History of Flour Milling in Early America online, accessed April 16, 2016; http://

www.angelfire.com/journal/millrestoration/history.html.

History Cooperative online, "The History of Divorce Law in the USA," accessed June 11, 2106; http://historycooperative.org/the-history-of-divorce-law-in-the-usa/.

The History of Loudoun County, Virginia online, "Early 19th Century Milling and Wheat Farming," accessed June 11, 2016; http://www.loudounhistory.org/history/agriculture-mills-and-wheat.htm.

The Homeroom online, E. Graham Alston, Inspector General of Schools. "*The Government Gazette*, May 28, 1870," accessed December 16, 2106; https://www2.viu.ca/homeroom/content/topics/statutes/rules70.htm.

Hoover & Truman, a presidential friendship online, "Part II: Feeding the World," accessed June 11, 2016; http://www.trumanlibrary.org/hoover/world.htm.

Indian Country Today Media Network online, "The War of 1812 Could Have Been the War of Indian Independence," accessed March 2, 2016; http://indiancountrytodaymedianetwork.com/2012/06/18/war-1812-could-have-been-war-indian-independence-11885.

iupui.edu online, "A Brief History of the Abolitionist Movement," accessed December 13, 2016; http://americanabolitionist.liberalarts.iupui.edu.

Jensen, Shirlene L. Rootsweb online "Aaron Lane Family Bible Records of the Little Miami Valley Hamilton County, Ohio," accessed May 25, 2015; http://rootsweb.ancestry.com/~ohhamilt/biblelane.htm.

McKivigan, John R., and Mary O'Brien Gibson. American Abolitionism online, "A Brief History of the American Abolitionist Movement," accessed April 2, 2016; http://americanabolitionist.liberalarts.iupui.edu/brief.htm.

Meek, Caroline. "On Which, Perhaps, a Mill Seat: The Mill and Water Power," *Hanford Mills Museum Newsletter*, Vol. 5, No. 2 Spring 1991, accessed May 18, 2016; http://www.hanfordmills.org/wp-content/uploads/2012/07/spring-91.pdf.

Mercy Wheat Kickoff April 1946, newsreel, YouTube, accessed June 11, 2016; https://www.youtube.com/watch?v=XGVzwr57R-8

Mill Creek Valley Conservancy District online, accessed December 14, 2016; https://sites.google.com/site/millcreekvcd/.

Mill Creek Watershed Council of Communities online, "Historical Timeline," accessed December 14, 2106; http://millcreekwatershed.org.

Mills and Wheat Farming in Loudon County Virginia online, "Early 19-Century Milling and Wheat Farming," accessed May 19, 2016; http://www.loudounhistory.org/history/agriculture-mills-and-wheat.htm.

National Association of Wheat Growers online, "Fast Facts," accessed December 14, 2016; www.wheatworld.org/wheat-info/fast-facts/.

National Library of Australia online, "Anti-Slavery Movement in the United States" accessed April 2, 2016; https://www.nla.gov.au/selected-library-collections/anti-slavery-movement-in-the-united-states.

Nebraska Studies online, "Pioneer Children: School," accessed June 11, 2016; http://www.nebraskastudies.org/0500/frameset_reset.html?http://www.nebraskastudies.org/0500/stories/0501_0207.html.

New Madrid, Missouri online, "Strange Happenings During the Earthquakes," accessed December 1, 2016; http://www.new-madrid.mo.us/index.aspx?nid=132.

New York Times online, Christopher Phillips, "The Breadbasket of the Union," accessed April 27, 2016; http://opinionator.blogs.nytimes.com/2012/08/08/the-breadbasket-of-the-union/.

Newspapers.com

Old Sturbridge Village online, "Lesson Plans: Farm Family," accessed April 27, 2016; http://resources.osv.org/school/lesson_plans/ howLessons.php ?PageID=R&LessonID=34&DocID=2037&UnitID=

Ohio History Central online, http://ohiohistorycentral.org.

Ohio Memory online, "Literary Ohio," accessed October 20, 2016; http://www.ohiohistoryhost.org/ohiomemory/wp-content/uploads/2014/12/TopicEssay_Literary.pdf.

Old Sturbridge Village online "Historical Background on the 19th-Century New England Farm Family," accessed April 2, 2016; http://resources.osv.org/school/lesson_plans/ShowLessons.php?PageID=R&LessonID=34&DocID=2037&UnitID.

Primitive Baptist online, "The Life of Wilson Thompson," accessed April 14, 2016; http://www.primitivebaptist.org/index.php?option=com_content&task=view&id=1405&Itemid=70.

Railroad Museum of Pennsylvania online "Trailways, Railways, & Roadways," accessed April 2, 2016; http://www.rrmuseumpa.org/education/index.shtml#resources.

Regional History from the National Archives online, "The Influenza Epidemic of 1918," accessed May 16, 2015; http://www.archives.gov/exhibits/influenza/epidemic.

RootsWeb's WorldConnect Project online, *Family Tree of Christopher W. Lane,* accessed March 9, 2015; http://wc.rootsweb.ancestry.com/cgi-bin/igm.cgi?op=GET&db=clane&id=I580.

Scholastic Facts for Now online, "Overland Travel Around 1800," accessed February 17, 2016;

http://factsfornow.scholastic.com/article?product_id=nbk&type=0ta&uid=10676862&id=a2022130-h.

Six Acres Bed and Breakfast online, accessed June 5, 2016; http://www.sixacresbb.com.

Slavery in the North online, "Race in Ohio," accessed April 2, 2016; http://slavenorth.com/ohio.htm.

Snopes.com, "1872 Rules For Teachers," accessed December 6, 2016; http://snopes.com/language/document/1872rule.asp.

Steffy, D Mylar. The Disciples Celebrationist online, "WWDBHS: What would David Burnet Have Said?" accessed May 3, 2016; https://disciplecelebrationist.com/2013/09/20/wwdbhs-what-would-david-burnet-have-said/.

Terrace Park Historical Society online, "The History of Avoca Park," accessed June 11, 2016; http://tphistoricalsociety.org/the-history-of-avoca-park/.

Tuberculosis History in Hamilton County online, accessed June 5, 2016; https://www.youtube.com/watch?v=Q8K34DSzaD8

Understanding Township and Range online, *Excerpts from Analysis of the System of United States Land Surveys in Standard Atlas of Knox County IL 1903*. George A. Ogle & Co. Chicago, 1903. Accessed May 24, 2016; http://publications.newberry.org/k12maps/module_06/images/township_range.pdf.

University of Michigan Library online, "Cincinnati, Ohio and the 1918-1919 Influenza Epidemic," accessed May 15, 2015; http://www.influenzaarchive.org/cities/city-cincinnati.html#.

U. S. Army Corps of Engineers, Louisville District, "Mill Creek, Ohio Flood Damage Control Project," accessed June 11, 2016; http://www.lrl.usace.army.mil/Portals/64/docs/ReviewPlans/Review%20Plan%20(Approved)%20-%20Mill%20Creek,%20OH%202-20-14%20without%20names.pdf.

U. S. Department of Health, Education, and Welfare, *100 Years of Marriage and Divorce Statistics, 1867-1967*, accessed June 5, 2016; https://www.cdc.gov/nchs/data/series/sr_21/sr21_024.pdf.

U. S. Department of the Interior National Register of Historic Places Inventory —Nomination Form, accessed May 24, 2016; http://cdn.loc.gov/master/pnp/habshaer/oh/oh0400/oh0415/data/oh0415data.pdf.

Vermont Timber Works online, " accessed June 19, 2016; http://www.vermonttimberworks.com/our-work/timber-trusses/queen-post-truss/.

Victorian Women: The Gender of Oppression online, accessed June 5, 2016; http://webpage.pace.edu/nreagin/tempmotherhood/fall2003/3/index.html.

Wikipedia.org.

Wilsey, Ashley M. "Half in Love with Easeful Death: Tuberculosis in Literature," (2012). Humanities Paper 11, accessed May 31, 2016; http://commons.pacificu.edu/cgi/viewcontent.cgi?article=1010&context=cashu.

Wirth, Jennifer. All Day online, "In the Victorian Era Tuberculosis Actually Inspired These Beauty Trends," accessed June 5, 2016; http://allday.com/post/8457-in-the-victorian-era-tuberculosis-actually-inspired-these-beauty-trends/?exp=3&utm_source=ADS&utm_medium=FBO&utm_campaign=HIP.

Worrall, Simon. National Geographic online, "Tracks to Freedom: The Inspiring Story of the Underground Railroad," accessed February 14, 2016; http://news.nationalgeographic.com/news/2015/02/150218-underground-railroad-slavery-civil-war-ngbooktalk/.

Young, Patrick, Esq. Long Island Wins online: "Immigrant America on the Eve of the Civil War," accessed on April 17, 2016; http://www.longislandwins.com/news/detail/immigrant_america_on_the_eve_of_the_civil_war.

Index

Index